The Strengths and Weaknesses of Wired and Wireless Communication

Mackil

First Printing, 2024

Contents

Chapter 1

Introduction

The use of wireless communication is now a fundamental part of our lives. It is used in devices such as smartphone, earphones, computer, television, or even your running shoes. We want faster data rate and constant connection without interruption. A higher frequency increase the possible data rate but also introduce a greater weakness to atmospheric effects such as rain, clouds, and atmospheric gases.

Wireless communication is electromagnetic radiation (EMR) that is transmitted from a source. As this radiation travels through the path it will interact with the space in various ways highly dependant on the frequency. This is due to the fundamental characteristics of wave propagation and how it interacts with the environment. A radio wave with low frequency and large wavelength is reflected by both the surface of Earth and the ionosphere i.e. specular reflection. In contrast a high frequency radio wave can have wavelengths similar in dimension to the roughness of the surface e.g. dirt, gravel, sand, or grass. The radio wave will be reflected in all possible direction and be lost i.e. diffuse reflection.

In the same way will the small size of the particles that make up our atmosphere become a larger problem with higher frequency. High frequency communication such as 5G, is interesting since it increase the possible bandwidth and more data can be transmitted in the same amount of time. However, the increase of data rate has to surpass the possible decrease from attenuation caused by atmospheric effects. A greater knowledge in the field of high frequency communication will open up faster communication with better mitigation of attenuation effects.

1.1 book objectives

The book study the signal from an Earth-space link and investigates the transmission impairments by statistical analysis. This book aims to:

- Summarise the most relevant atmospheric effects on the Earth-space radio channel. The focus will be on the millimeter wavelength which is relevant for the signal transmitted by the Alphasat satellite.

- Overview the effects of the air water vapor contents on the millimeter wavelength propagation. A focus will be on water vapor and clouds due to broad field of attenuation effects. This will help in understanding the data in the section results and discussion.
- Analyse the correlation between the measured received signal level and the air relative humidity and temperature measurements. The correlation function measures the similarity of two signals and will be used to find if there is any relationship.
- Compare measured attenuation and estimated attenuation by ITU methods of cloud and atmospheric gas attenuation calculated using the surface meteorological data.

1.2 Radio wave propagation

1.2.1 Ionospheric interference

Communication between a ground station on Earth and a satellite in space can be hindered by the presence of the ionosphere. This barrier will act as a reflector or absorber for frequencies below about 30 MHz[1]. Ionosphere is the upper layer in Earth's atmosphere (altitude roughly between 15 km to 400 km) where particles are ionized by the solar wind. The solar wind changes in activity by both in relation to the solar cycle and seemingly random out-burst such as coronal mass ejections. The ionized particles will vary also by the diurnal cycle as the electrons decrease during night time. This is determined by the plasma frequency ω_p of the ionopshere and the transmitted frequency ω by the radio wave. It can simply be said that a larger plasma frequency will both refract and absorb the incoming radio wave. A larger transmitted frequency will be less affected by the ionosphere until a point when it is essentially transparent.

1.2.2 Terrestrial communication

Below 30 MHz space communication is not possible and there are different modes of operating. In general the radio wave will follow the surface of Earth guided between the ionosphere and surface.
For frequencies up to 300 MHz the propagation can bounce on different levels

in the ionosphere and be refracted back towards the surface of Earth known as ionospheric or sky wave.
Up to frequencies of around 3 GHz the ionosphere can be used to scatter the radio wave due to the irregularities of the refractive index in the troposphere[1]. Space communication is also viable as the frequency penetration limit of the ionosphere is reached. Satellites in low earth orbit often operate in this,region. For use in terrestrial communication it can be unreliable due to the dynamic nature of the ionosphere.

1.2.3 Earth-space communication

For Earth-space communication a frequency well beyond the penetration limit of 3 GHz or more will ensure minimal interference with the ionosphere. Radio waves in this frequency range will travel unimpeded until it reaches absorption line frequencies of atmospheric gases. The only two gases that have a significant effect on the signal studied in this book are oxygen and water vapor. The signal will also become more vulnerable to effects such as rain and clouds.

1.3 Alphasat and Aldo Paraboni Payload

Alphasat is the largest telecommunication satellite in Europe. It was launched on the 25th of July 2013 and carries the Aldo Paraboni Payload (in honor of Italian scientist Aldo Paraboni) which transmits two coherent beacon signals of 19,702 GHz and 39,402 GHz. The satellite is in geosynchronous orbit at longitude 25° east and with a maximum elevation angle of ±3° from the equator[2]. This orbit essentially matches the angular velocity of Earth which causes the satellite to remain practically motionless for an observer on the surface.
The Aldo Paraboni Payload investigates the use of higher frequencies in Q- and V-band for telecommunication as it both provide more bandwidth and avoid congestion of the currently used Ku/Ka band. As we move up in the higher frequency bands the signal becomes more susceptible to attenuation of atmospheric effects. Turbulent atmospheric phenomenon such as thunder-and rainstorms can cause weakening up to several tens of dB. More com-monly occurring effects such as cloud cover and atmospheric gases can also cause significant degradation. One of the goals of the Alphasat mission is

to characterise and further develop prediction methods to understand these diminishing effects while gaining the benefits of greater bandwidth.

1.4 Ground station in Budapest

Budapest University of Technology and Economics (BME) operates three antennas receiving transmission from Alphasat. The three antennas located on top of the university building in Budapest and have a radius of 180 cm, 60 cm, and 30 cm. There is also gathering of meteorological data such as rain rate, temperature, relative humidity, and wind parameters (see Table 1.2). Budapest, Hungary is situated in the middle of the Carpathian Basin, a plain land covered by the Carpathian mountain and the Alps. The receiver station is placed on the building of the Department of Broadband Infocommunications and Electromagnetic Theory, 120 meters above sea level with coordinates 47,48°N latitude and 19,06°E longitude[3]. A visual representation of the receiver station and Alphasat satellite is seen in Figure 1.1.

	Latitude	Longitude
Receiver station	47,48°N	19,06°E
Alphasat	±3°N/S	25°E

Table 1.1: Coordinates of the receiver station and transmitting satellite.

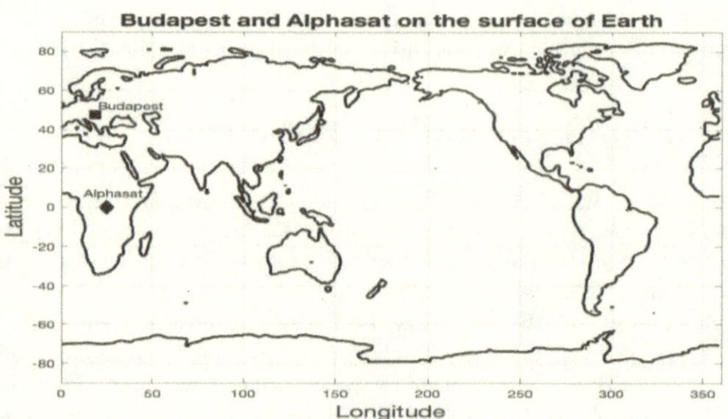

Figure 1.1: Receiver station in Budapest marked with a square and the transmitting satellite Alphasat marked with a diamond on their relative position on the surface of Earth.

Measured parameter	Measured Unit	Resolution
Attenuation	dB	1 Sample/sec
Rain rate dropcounter	mm/hr	1 Sample/min
Rain rate tipping bucket	mm/hr	1 Sample/min
Temperature	C	1 Sample/min
Relative humidity	%	1 Sample/min
Wind speed	m/s	1 Sample/min
Wind gust	m/s	1 Sample/min
Wind direction	m/s	1 Sample/min

Table 1.2: Measured data conducted by BME.

1.4.1 Elevation angle

The antennas require an elevation angle to accurately point to the satellite in geosynchronous orbit. The elevation angle is important to consider in prediction of attenuation. A small angle (with reference to the horizon) will encounter more atmosphere on the path between satellite and antenna. The mean elevation angle of the BME antennas are 35°[3].

	Frequency (GHz)	Diameter (m)	Mean elevation angle
Antenna 1	19,702	0.6	35°
Antenna 2	39,402	0.3	35°
Antenna 3	39,402	1.8	35°

Table 1.3: Receiving antennas operated by BME.

1.4.2 Tracking system

Alphasat is a moving object in the sky which will have variances due to orbital parameters. Both the elevation angle and azimuth can change slightly. Even a small drift out of the antennas direction can cause a loss of power. This is accounted for by a program tracking system. This continuously calculates the position of Alphasat and properly adjust the angle of the antenna. This adjustment is done periodically and can be visible on the signal (see Figure 4.24). This pointing error is kept below +/- 0,5 dB.

Chapter 2

Background

2.1 History

The earliest form of communication by electrical means was with transmission lines connecting a source to a receiver. This was first done around mid 1800s with the invention of telegraphy. The foundation of EMR laid by Maxwell equations showed it was theoretically possible to create wireless communication without the need of transmission lines. This was not commercially used until 1897, 43 years after Maxwell's discovery. The foundation of electromagnetic theory led Marconi to the first wireless telegraphy patent[4]. A decade later the vacuum tube amplifier and oscillator made it possible to transmit voice and the development of telecommunication rapidly increased from that point[4]. The two different ways of communicating have their own strengths and weaknesses depending on which environment can be found between the source and receiver. In a large city with tall buildings, short distances, and possibly other obstacles a wire system is much more reliable. For long distances or difficult terrain, a wireless system may be deemed more efficient.

2.2 Electromagnetic radiation

EMR will propagate with a definite speed of light c in vacuum and spread outwards in space. It is a wave of photons and unlike mechanical waves i.e. sound, water, it does not require a medium to propagate. The energy E is defined by Planck's constant h and its frequency f while the wave is defined by its wavelength λ and frequency f. The frequency of an electromagnetic wave will not change while passing into different mediums[5]. The photon/wave velocity v is only the speed of light in vacuum and can change depending on the index of refraction n of the medium. As an EM wave travels into a denser medium such as Earth's atmosphere the velocity will be slightly decreased. This is an important aspect since wireless communication can be delayed if the path have large refractive index irregularities which can be observed in temperature changes.

$$E = hf \quad [J] \tag{2.1}$$
$$v = \lambda f \quad [m/s] \tag{2.2}$$
$$v = c/n \quad [m/s] \tag{2.3}$$

2.2.1 Radio refractivity

Index of refraction for air $n_{air} = 1,0003$ is very close to the value of vacuum $n_{vacuum} = 1$. It is often more convenient to work with the radio refractivity N when dealing with radio wave propagation in the atmosphere. A medium have both a real N' and imaginary part N''. The real part describes the oscillating properties of the medium while the complex part describes the absorption. The refractivity index of a medium is dependant on the frequency. The knowledge of a mediums specific refractivity opens up the possibility to describe atmospheric effects on such as attenuation γ, phase change of the wave β, or a propagation delay τ.

$$N = (n - 1) * 10^6 \tag{2.4}$$
$$N = N'(f) + iN''(f) \tag{2.5}$$
$$\gamma = 0.1820 * f * N''(f) \quad [dB/km] \tag{2.6}$$
$$\beta = 1.2008 * f * N'(f) \quad [deg/km] \tag{2.7}$$
$$\tau = 3.3356 * N'(f) \quad [ps/km] \tag{2.8}$$

2.2.2 Properties of different mediums

There are essentially four different mediums that different space can be divided into. Free space, lossless dielectric, lossy dielectric, and good conductors. They are described by their conductivity σ, permittivity ϵ, and permeability μ. The conductivity of a material is the measurement how well electric current can propagate through it. It is the opposite of resistivity which is a measurement how good a material resist electric current. Permittivity measures the polarization of a material. A medium with large permittivity polarize to a larger degree. Vacuum has the value of $\epsilon_0 = 1$. Permeability measures the magnetisation of a material when affected by a magnetic field.

$$
\begin{aligned}
&Free\,space: & \sigma = 0, \epsilon = \epsilon_0, \mu = \mu_0 \quad (2.9)\\
&Lossless\,dielectric: & \sigma = 0, \epsilon = \epsilon_0\epsilon_r, \mu = \mu_0\mu_r \quad (2.10)\\
&Lossy\,dielectric: & \sigma \neq 0, \epsilon = \epsilon_0\epsilon_r, \mu = \mu_0\mu_r \quad (2.11)\\
&Good\,conductor: & \sigma \simeq \inf, \epsilon = \epsilon_0, \mu = \mu_0\mu_r \quad (2.12)
\end{aligned}
$$

2.2.3 Attenuating wave

A travelling wave in z-direction can be described by its initial amplitude E_0 and Euler's formula e^{-jkz}. k is the wave number and can be represented by both a real k' and imaginary part k''. The real part determines the oscillating characteristics while the imaginary part is the dampening or attenuation. It can mathematically be represented with attenuation coefficient α and phase constant β. This will show an oscillating wave with decreasing amplitude over time(see Figure 2.1).

$$\overline{E(z)} = E_0 e^{-\alpha z} e^{-j\beta z} \quad (2.13)$$

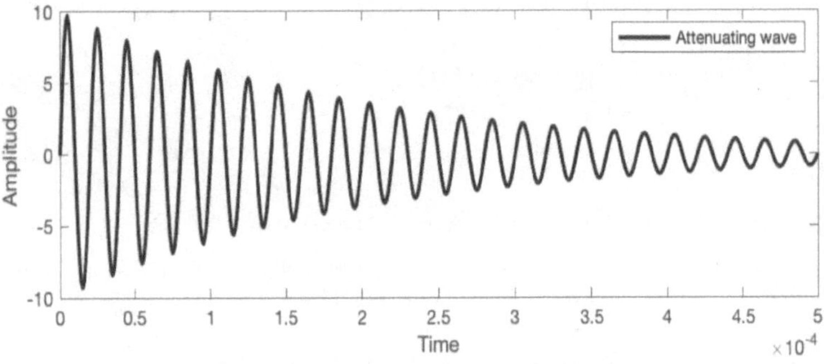

Figure 2.1: Example of an attenuating wave.

2.2.4 Attenuation, loss of signal power

Attenuation is often an undesirable product for all type of communication and the two means of communication have their own strength in this regard.

It can however be used to its advantage, e.g. short range communication without the concern of congestion from other sources. Transmission lines have an exponential decrease while radio waves attenuate with the inverse-square law[4]. Attenuation is commonly measured in dB/km and is highly dependant on the frequency of the signal.

2.2.5 Inverse-square law

The inverse-square law states that isotropic radiation distributes over a sphere at a radius r from the radiation source. It should be noted that antennas can be constructed such as it radiates in a more concentrated region. The radiation will however propagate in the same way regardless. The received intensity I_r at a distance R from the transmitted power source P_t is related by the mathematical expression below. The power received P_r is also given.

$$I_r = \frac{P_t}{4\pi R^2} \qquad [W/m^2] \tag{2.14}$$

$$P_r = P_t\frac{R_A^2}{R^2} \qquad [W] \tag{2.15}$$

This shows the dependency of the power and the distance squared. R_A is the antenna radius and show that a larger size can receive more power. It also shows that a larger transmitting antenna can radiate more power with size. When working with very large or small values (such as the study of attenuation) a logarithmic scale is useful to represent the values. The dB-scale is often used for attenuation. The attenuation be calculated in dB between two locations if the power of the signal is known.

$$A = 10 * log_{10}(\frac{P_1}{P_2}) \qquad [dB] \tag{2.16}$$

$$A = 20 * log_{10}(\frac{R_2}{R_1}) \qquad [dB] \tag{2.17}$$

P_1 represent the power received at location with distance R_1 while P_2 at a distance R_2. If the same size of antenna is used such as R_A is equal the only two parameters that will not cancel out is the distance from the source to the antenna.

2.2.6 Comparison of attenuation between a transmission line and radio wave

Assume a transmission line with a specific attenuation of 5 dB/km and length 20 km. This could apply to a coaxial transmission line with a maximum frequency up to a few MHz. This particular system would suffer an attenuation A_{TL} of 100 dB. Transmission lines over long distances are only feasible at lower frequencies below a few MHz. At higher frequencies the cable attenuation can reach tremendous loss of 10 dB/100 m or even far more. High frequency communication over long distances is thus a feature exclusive for radio communication. Assume a transmitting radio wave that propagate the same distance of 20 km with power P_{20} which suffer the same amount of attenuation of 100 dB. A doubling of the distance to 40 km will yield another 100 dB for the transmission line while only a decrease of 6 dB for the radio wave.

$$A_{TL} = 5 * x \qquad [dB] \tag{2.18}$$

$$A_{EM} = 10 * log_{10}(\frac{P_{20km}}{P_{40km}}) \qquad [dB] \tag{2.19}$$

$$A_{EM} = 20 * log_{10}(\frac{40}{20}) \approx 6 \qquad [dB] \tag{2.20}$$

Here x is representing the length and 5 is the specific attenuation for the transmission line. P_{20} and P_{40} represent the power received at 20 km and 40 km respectively. The attenuation give for the transmission line is easily shown by its linear relationship with km. For the radio wave, the fraction will depend on the two distances squared. A doubling of the distance will always yield the result $log_{10}(2)$ which multiplied by 20 shows why the attenuation is equal to 6 dB. If we consider that dB is a logarithmic scale, a decrease of 100 dB is a reduction of the signal power by a factor of 10^{-10}. It is clear that the use of transmission lines over long distances can cause much larger attenuation than radio waves. It should be noted that the radiated source have a very large coupling loss since the receiving antenna only covers a small region. However, transmission lines can accurately predict the amount of attenuation while radio waves can suffer from unpredicted effects. If the example given above only accounted for attenuation of a standardised atmosphere, i.e. a water vapor content of 7,5 g/m^3, unpredicted effects such as rain (increasing the humidity) or congestion from other sources of radiation could increase the attenuation even further.

2.3 Different modes of radio wave propagation

2.3.1 Terrestrial and Space communication

The frequency of the signal will determine both the characteristic of the wave and the impairments that can affect it. Wireless communication is operating from only a few Hz up to several GHz and include many different regions. A wide classification can be done for frequencies below and above the ionospheric penetration frequency of around 30 MHz. Terrestrial communication is operating below this frequency while the Earth-space radio channel require a frequency above. Another aspect of a preferred high frequency for satellite systems is the antenna size. Large antennas are not easy to bring into space and an antenna can not be much shorter than a quarter of the wavelength to radiate in an efficient way[4]. The upper layer of the atmosphere impose a barrier for frequencies below a certain threshold. This fluctuates depending on the activity of the Sun which induce this effect by its varying stream of radiation. The ionized electrons in the ionosphere oscillate with a certain frequency which determines the plasma frequency. Terrestrial communication below the plasma frequency will be trapped making it possible to communicating over the horizon by reflected or scattered radio waves. These type of radio waves are referred to as surface waves. The greatest impairment for these waves are the conductivity of the surface it travels over. The distance can be increased by communicating over oceans as water is a better conductor than the surface of Earth. Typical distances for this type of communication is usually a few hundred kilometers and is where AM broadcasting occur. A second type of communication in the MHz region up to 300 MHz is reflected by the ionosphere called sky waves. This is a region used by FM broadcast and television. The reflected sky waves enables communication beyond the horizon and can even reflect multiple times creating a world-wide wave. A third type of communication is in the region from MHz up to 3 GHz where the wave is scattered by the different layers of the ionosphere. The dynamic nature of the layer makes it susceptible to disturbance by solar radiation and some what unreliable. Far beyond the penetration frequency of the ionosphere above 3 GHz reliable space communication can be conducted. The largest impairment is atmospheric gases and hydrometeors such as rain, snow, or cloud. Most of the attenuation is caused by absorption but also by scattering. The atmospheric gases that have significant effect is oxygen and water vapor. Both of them have absorption lines at difference GHz frequencies. Certain regions are therefore best avoided since attenuation levels of several decibel per kilometer are achieved. The most significant being the family of absorption lines by oxygen around 60 GHz. The specific attenuation for atmospheric gases is illustrated in Figure 2.2. Full line represent dry air only composed of oxygen. Dotted line is both oxygen and water vapor with concentration of $7,5g/m^3$.

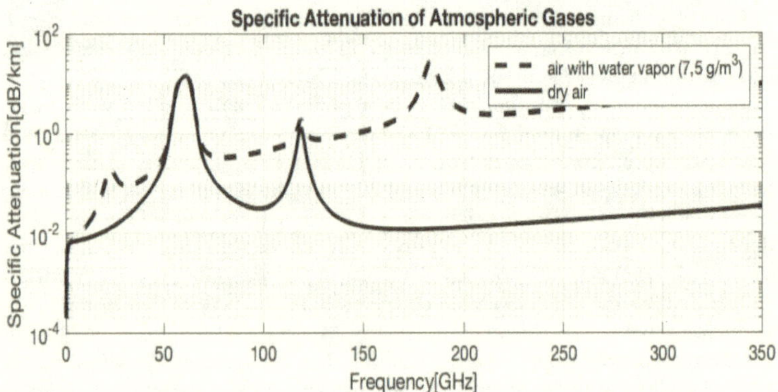

Figure 2.2: Specific attenuation by air constituted by oxygen and water vapor. Full line represent dry air (oxygen) while dotted line show oxygen and water vapor with concentration of 7,5 g/m^3. Figure was created in Matlab by ITU method explained in section 3.5.

2.4 Atmospheric impacts on the Earth-space radio channel

2.4.1 Rain

Rain have the largest detrimental effect on the signal for the millimeter wavelength. Most of the power is lost by the lossy dielectric of water droplets but there is also loss due to scattering. The absorption is however the dominant factor of attenuation and scattering is relative small[4]. The shape of the rain drops can change into non-spherical geometry. This can also cause polarization of the signal. Rain events can also differ by different structures depending on cloud type. There are especially two types that are important for earth-space communication, stratiform and convective.[1]. The large scale stratiform rain that can cover hundred of kilometers in horizontal distance. These are typical longer period events with evenly spread distribution of rain rate. Convective rain come from cloud with large vertical height and shorter spatial distance. The rain can come with shorter burst with higher rain rates and also includes thunderstorms. Even though rain potentially cause the largest impairment on the signal the temporal distribution is relative short. This will differ depending on geographical location and different type of diurnal, seasonal, and yearly variations.
The modelling of rain attenuation is a difficult task since the rain parameters include drop shape, distribution, rain rate, frequency, and to some degree temperature and pressure[1]. These parameters will also change over the path and would require tremendous power to accurately model. The most practical models use the rain rate. The relationship of rain rate and attenuation is derived from analytical analysis. The current models have shown enough accuracy for use by communication engineers[6].

2.4.2 Cloud and fog

Clouds can also cause a large detrimental effect and have a higher probability of occurrence than rain (40-80%)[7]. The effects of clouds can also last for long periods of time, potentially causing a much more long term impairment. As the air mass reaches saturated or supersaturated levels of water vapour (100% relative humidity) it can condense into clouds and fog. Clouds are not water vapor as the particles made by the cloud are small water droplets of less than 0.1 mm in diameter[1]. On the millimeter wavelength, the liquid water content has to rise above $0.001 g/m^3$ to reach significant attenuation levels[8]. Fog can be thought as a very light condensed matter of water vapor and is identified by a mass content per unit of air of 0.001-0.5 g/m^3[8]. Typical medium fog with a visibility of 300 meter has a value of 0.05 g/m^3 while thick fog with a visibility of 50 meter reach values up to 0.5 g/m^3[9]. Non-precipitating clouds contain a concentration levels of 0.1-1 g/m^3 while rain clouds can exceed 2 g/m^3[8]. The peak rate of maximum liquid water has been measured up to 5 g/m^3 for thunderstorm clouds[1]. Fog might not be significant until above the frequency of 100 GHz because of its low water content and small area effect.

2.4.3 Atmospheric gases

The atmospheric gases, oxygen and water vapour will have least impact for most regions on the spectrum except if the signal is operating close to the absorption lines. The dry air referencing the impact of oxygen is assumed constant over the atmosphere while the water vapor content can fluctuate greatly and is measured by its humidity.

Earth atmosphere composition

The atmosphere contains many species but can be said to be made up by mainly nitrogen (78 %) and oxygen (21 %) with small contribution of argon (0.9 %), carbon dioxide (0.1 %), water vapor (variable %) and trace gases.

Most dominant species: oxygen and water vapor

Even though the atmosphere is made up by a majority of nitrogen it does not contribute to the attenuation. This is because of its diatomic structure which only stretch or compress. That is, the molecule can not have a rotational motion which would create an unbalance in the symmetry and a dipole moment which could interact with incoming signals. Water vapor is a polyatomic molecule with possible normal modes which give rise to rotational vibrations that create an electric dipole moment. Oxygen despite being diatomic is relevant in the microwave region since it possess permanent magnetic dipole moment and with its high abundance in the atmosphere it still create a potentially high attenuation[10]. Water vapor has a relative small abundance compared to oxygen and nitrogen as they both make up 99 percent of the atmosphere together. Water vapor also has the characteristic of being highly variable depending on time of the day, geographical position, and weather activity[11]. The other gases stay relative constant through out the atmosphere and can be modelled with an exponential decrease with height[11]. For frequencies below 1000 GHz water vapor and oxygen are the only gases that will make a significant decrease on the signal[7]. That is, no other gas have absorption lines with considerate effect. The absorption lines for water vapor occur at 22 GHz and 180 GHz. Oxygen has many absorption lines creating a family of absorption bands at 60 GHz and a single line at 120 GHz. The atmosphere is opaque in the 60 GHz region with attenuation levels exceeding 20 dB making it a region that is avoided for communication[11]. This can however be used for shortwave range communication where the large attenuation allow a reuse without fear of congesting other sources.

2.4.4 Scintillation

Scintillation are fast fluctuations in the signal which can be caused by both the electron density irregularities in the ionosphere and the irregularities of the refractive index in the troposphere. For frequencies above 6 GHz the ionosphere impose no impairment on the signal and the scintillation effect is caused by humidity and temperature changes in the troposphere. Earth-space signals up to 50 GHz have shown affected[1]. The degree of impairment depends largely on elevation angle as more of the path is obstructed by the troposphere. Below 10°elevation angle the effect can increase to dramatic levels, to devastating quantity similar to rain. The effect is increase during summer month and by the presence of clouds.

2.4.5 Radio noise

Electromagnetic radiation in thermodynamic equilibrium will emit the same radiation as it absorbs by Kirchoff's law. That means that any gaseous absorption or rain absorption will emit the same radiation and disrupt the signal with noise. Examples of sources of radio noise include the cosmic background noise, other communication system, solar- and lunar radiation.

2.4.6 Elevation angle

Where the receiving ground station (antenna) is pointing into space is also of great importance to consider for the attenuation. If more atmosphere is travelled through on the path length more of the signal will interact with the medium resulting in larger attenuation. It will decrease with increasing elevation angle as the shortest path through the atmosphere is at zenith (90°elevation angle).

Chapter 3

Method

3.1 Methods of predicting attenuation

To perfectly predict attenuation on a path length requires perfect knowledge of the environment. The amount of particles, the dynamics of the weather and all seemingly random effects that occur in the space is impossible to account for perfectly. There are instruments that can give good estimates of the atmosphere such as radiometers measuring the radiation or radiosondes that measure the vertical structure of a local area. The amount of computational power it would require to model every effect is beyond today's technology and thus is statistical analysis the best method to predict impairments on a signal. There are a vast amount of different models to predict the atmospheric effects ranging in complexity and practicality. This is due to the variance in different location and the variance of the climate on the planet with no one-fit-all method.

3.1.1 International Telecommunication Union (ITU)

ITU make extensive work in prediction methods and give both complex accurate methods and simple estimation methods. ITU has descriptive approximate methods that are easily accessible from their website. These methods are derived from data gathered from decades of studying the atmosphere. Iterations of the methods are constantly being worked upon and improved. There are also documentation of standard reference atmosphere for calculations without the need of any locally gathered meteorological data.

All of the methods utilised in this book are from the ITU database as recommended in documentation [6].

3.1.2 Comparison between prediction methods and measured data

The prediction of attenuation by cloud and fog, and atmospheric gas are compared with the locally measured attenuation gathered by BME. Atmospheric gas can be estimated by local meteorological data while cloud and fog uses theoretical values as recommend in [9].

3.2 Interpolation of meteorological data and coordinates

During my work I applied the meteorological and received signal data recorded in 2018 by the Alphasat receiver station in Budapest. Path attenuation can be calculated by subtracting the momentary received signal power from the clear sky level.

The resolution of the meteorological data and the attenuation are not equal as seen in Table 1.3. The meteorological data is measured each minute while the attenuation every second. There are 30 days in April which provides 43200 meteorological data points and 2592000 attenuation data points. Each of the 43200 points were interpolated over 2592000 data points to match accordingly. Each meteorological data point is thus extended over 60 data points (see Figure 3.1. This made calculation with cross-correlation possible as both data set now match together.

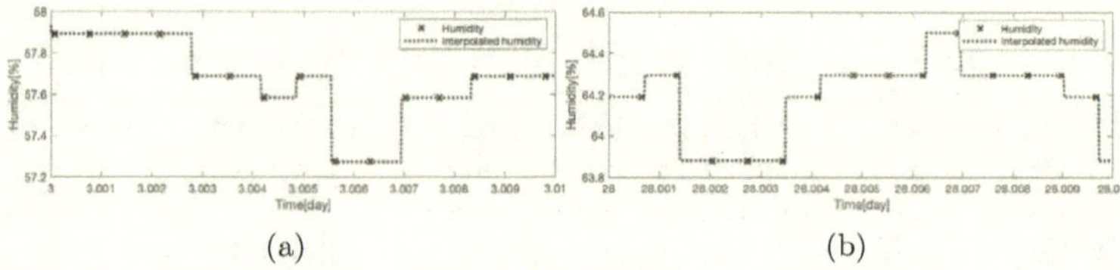

(a) (b)

Figure 3.1: Showcase of how the interpolation method changes the data points for a lower and upper part of the spectrum. A difference can be seen in how the data points extend to larger values for the lower part (Fig. (a)) and to lower values in the upper part (Fig. (b)).

3.3 Cross-correlation/Auto-correlation

The similarity of two signals can be measured by cross-correlation. It is a great tool in signal processing to scan for a short signal in a longer one. In our case the meteorological data will be compared with attenuation and with itself. One of the signals can be visualized as fixed in time while the other signal is shifted over the signal length. It can thus be calculated where on the signal the best match is found. Since the shifted signal will differ in length a zero-pad is done over the missing region. The shift is represented by the variable τ where the length is equal to two times the vector length subtracted by one (to avoid counting the 0 twice). If the same function is chosen for both f(t) and g(t) as shown in Equation (3.1) below it is known as auto-correlation. This can be used to find repeating patterns in a signal.

For two discrete signals the product of each data point are summed. If both values at a point show large positive or negative value it will contribute to the sum with a large amount. If the signal value show opposite sign a negative value will be added. As this process is done for every possible position of the shifted signal a new function is created, the correlation function. The mathematical process in shifting a signal over another is known as convolution. However, cross-correlation and convolution differ as one of the function are reversed and flipped in the latter case. These operations are also related to Fourier transform. The fastest computational way to get the cross-correlation function is to take the inverse-Fourier transform of the product of the Fourier transform of both function. This is known as the convolution theorem as seen in Equation (3.2).

$$R(\tau) = \int_{-\infty}^{\infty} f(\tau)g(t-\tau)d\tau \tag{3.1}$$

$$R(\tau) = \mathcal{F}^{-1}\left(\mathcal{F}[f] * \mathcal{F}[g]\right) \tag{3.2}$$

The use of cross-correlation will be of great use for interpretation of data in the results section. An important note are that the signals are subtracted by its first vector value. This will make sure the two compared signals both start at 0 on the y-axis. If this step is skipped the cross-correlation will only look for highest positive correlation (since both humidity and temperature is positive values). We thus measure the change of the meteorological data over the chosen intervals. This is visible by the Δ-symbol on the axes. Example of calculating cross-correlation without subtraction (left column) and with subtraction (right column) is shown in Figure 4.4.

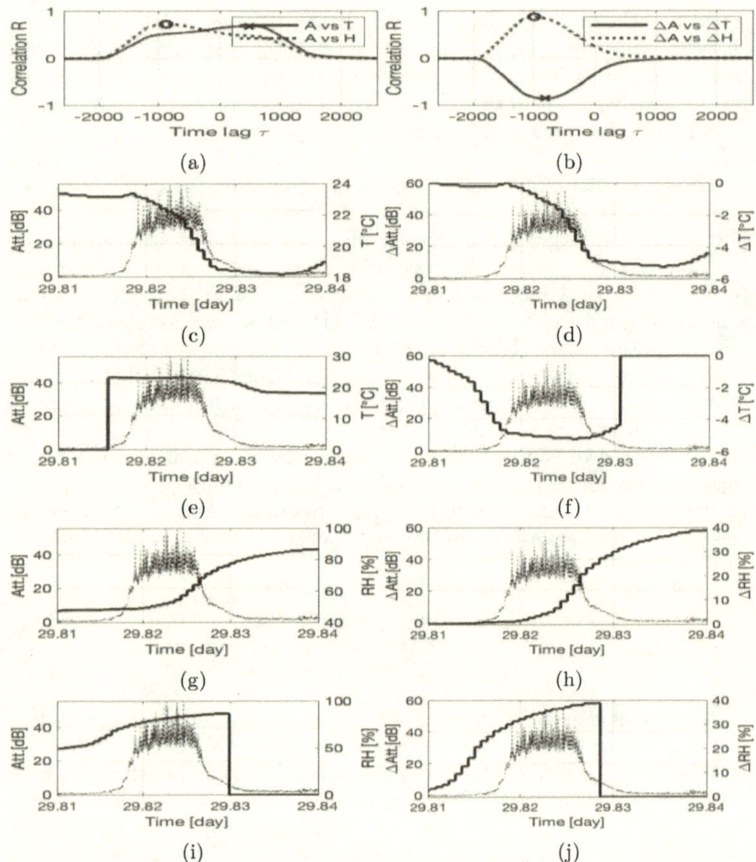

Figure 3.2: Comparison of calculation of cross-correlation. Figures (a) and (b) show the cross-correlation function over time lag τ. Other figures show both actual time and the shift of meteorological data where the maximum correlation was found.

3.4 Clear-sky level

Radio waves will attenuate naturally by itself due to the inverse-square law. This attenuation is not interesting in the research of other attenuation effects. There are different ways of predicting the amount of attenuation caused by propagation in free space, a simple method is to take the mean of the measured received power and subtract the momentary received power (see Figure 3.3b. This will include attenuation from all type of effects and is not the most accurate method. However, it gives a rough estimate to work with and is applied for the data in this book. Example of another more accurate method is to use a radiometer that can accurately measure the clear-sky level. This was not available for the BME data set used. The BME data from April 2018 show a clear-sky level of -52.7 dB. Region with larger attenuation than the mean can now be studied and assumed to be caused by atmospheric effects.

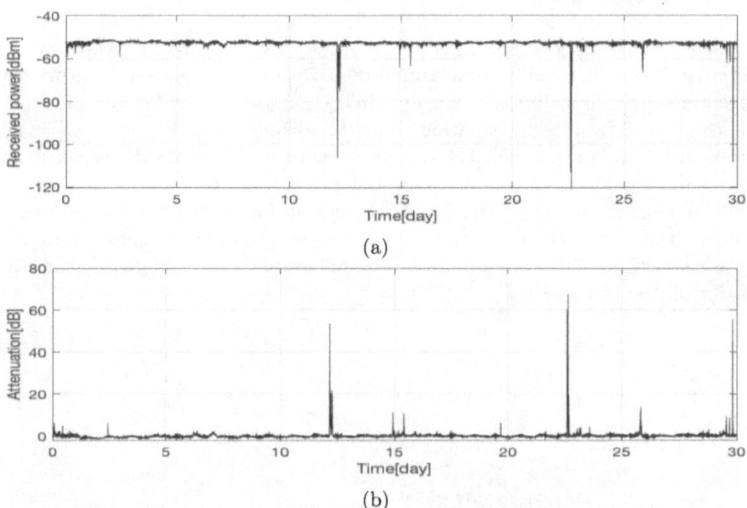

Figure 3.3: Raw data of received attenuation (Fig. (a)) and adjusted attenuation with subtracted mean (Fig. (b)), April 2018.

3.5 Method of predicting attenuation by atmospheric gases

The attenuation from atmospheric gases can be predicted from ITU recommendations [12]. There are two methods to chose from: by using local radiosonde measurements over the vertical column or by approximation with local surface measurements. Since the BME local meteorological data available is limited to surface parameters the approximation method was chosen. It is accurate for signals up to 350 GHz. A detailed description is found in Annex 2 in [12]. The method requires three parameters:

- dry pressure, p [hPa]
- temperature, T [°C]
- water vapor density, ρ $[g/m^3]$

Only the temperature is directly measured by BME while the water vapor density can be derived from measured relative humidity and temperature by recommendation by Ippolito, page 142 in [1] Equation (7.5). The dry pressure is calculated by recommendation from [13] (Equation (3.3)) since no local pressure data is available. The dry pressure is derived from the barometric pressure B and the water vapor partial pressure e (Equation (3.4)). The ITU document [13] give recommended equations for both winter and summer month which are dependant on the height above ground. The summer model was chosen for this case but could be debatable since month of April is between the two seasons.

$$B(h) = 1008.0278 - 113.2494h + 3.9408h^2 \qquad [hPa] \tag{3.3}$$

$$B = e + p \qquad [hPa] \tag{3.4}$$

$$e = \frac{\rho T}{216,7} \qquad [hPa] \tag{3.5}$$

$$p = B - e \qquad [hPa] \tag{3.6}$$

$$\rho = \frac{rh}{5.572} * \theta^6 * 10^{(10-9.834*\theta)} \qquad [g/m^3] \tag{3.7}$$

$$\theta = 300/(T + 273.15) \qquad [K^{-1}] \tag{3.8}$$

3.5.1 Specific attenuation

The specific attenuation in units [dB/km] is calculated by the frequency dependant complex refractivity. The refractivity is calculated from summation by absorption lines of oxygen and water vapor. The method uses 44 lines of oxygen and 35 lines of water vapor.

$$\gamma = \gamma_o + \gamma_w = 01820 * [N''_{Oxygen}(f) + N''_{WaterVapor}(f)] \tag{3.9}$$

$$N''_{Oxygen}(f) = \sum_{i(Oxygen)} S_i * F_i + N''(f)_D \tag{3.10}$$

$$N''_{WaterVapor}(f) = \sum_{i(WaterVapour)} S_i * F_i \tag{3.11}$$

The complex refractivity is given by the strength of the oxygen or water vapor line, S_i and the respectively line shape factor F_i. These are found in tables 1 and 2 in ITU document [12]. $N''(f)_D$ is the dry continuum due to pressure-induced nitrogen absorption and the Debye spectrum.

$$S_{i(Oxygen)} = a_1 * 10^{-7} * p * \theta^3 * exp[a_2(1-\theta)] \tag{3.12}$$

$$S_{i(WaterVapor)} = b_1 * 10^{-1} * e * \theta^{3.5} * exp[b_2(1-\theta)] \tag{3.13}$$

$$F_i = \frac{f}{f_i}\left[\frac{\Delta f - \delta(f_i - f)}{(f_i - f)^2 + \Delta f^2} + \frac{\Delta f - \delta(f_i + f)}{(f_i + f)^2 + \Delta f^2}\right] \tag{3.14}$$

$$N''(f)_D = fp\theta^2 \left[\left(\frac{6.14 * 10^{-5}}{d\left[1 + (\frac{f}{d})^2\right]}\right) + \left(\frac{1.4 * 10^{-12} p\theta^{1.5}}{1 + 1.9 * 10^{-5} f^{1.5}}\right)\right] \tag{3.15}$$

f_i is oxygen or water vapour line frequency corresponding to the absorption line, Δf is the corresponding width of the line. The width of the line has to be modified to account for Zeeman splitting for oxygen and Doppler broadening for water vapour. There also has to be a correction factor δ included for the oxygen due to interference effects. d is the width parameter of the Debye spectrum. f is the frequency of the operating signal which can be between 1-350 GHz.

$$\Delta f_{Oxygen} = a_3 * 10^{-4}(p * \theta^{0.8-a_4} + 1.1 * e * \theta) \tag{3.16}$$

$$\Delta f_{Oxygen} = \sqrt{\Delta f^2 + 2.25 * 10^{-6}} \tag{3.17}$$

$$\Delta f_{WaterVapor} = b_3 * 10^{-4}(p * \theta^{b_4} + b_5 * e * \theta^{b_6}) \tag{3.18}$$

$$\Delta f_{WaterVapor} = 0.535\Delta f + \sqrt{0.217\Delta f^2 + \frac{2.1316 * 10^{-12} * f_i^2}{\theta}} \tag{3.19}$$

$$\delta = a(_5 + a_6 * \theta) * 10^{-4} * (p + e) * \theta^{0.8} \tag{3.20}$$

$$d = 5.6 * 10^{-4}\,(p + e)\,\theta^{0.8} \tag{3.21}$$

This concludes the method to calculate the specific attenuation due to the atmospheric gases by oxygen and water vapor. The spectroscopic data parameters a and b are found in tables in ITU document. Image of the calculated specific attenuation is shown in figure 2.2 where the water vapor density was set to a fixed value of 7,5 g/m^3.

The calculated specific attenuation is a parameter that will change depending on the involved variables over the vertical column. Thus the specific attenuation of gases will change as pressure drops, temperature changes, and water vapor density changes. For the case of local radiosonde data these parameters can be gathered over different layers of the atmosphere. For example, a moving air mass with a larger concentration of water vapor will not cover the whole path but perhaps only a few hundred meters up to few kilometers. Thus the parameter of this air mass has to be known for proper estimation. In our case where only the surface parameters are available the integrated path can not be calculated. The method only uses a single specific attenuation for the surface level and estimates an equivalent height which estimates the total zenith attenuation without integration. The calculation is done for zenith angle and is then adjusted for the proper elevation angle by the cosecant law.

3.5.2 Equivalent height

The equivalent height is an approximation of what height is multiplied with the corresponding specific attenuation based on the surface parameters. This method assume that the specific attenuation have an exponential decay versus altitude. Equivalent height for oxygen h_o and water vapor h_w are calculated separately to match their corresponding specific attenuation. There is one constraint that $h_o < 10.7r_p^{0.3}$ when $f < 70GHz$. T is the temperature at the surface of Earth, ρ is the water vapor density at the surface, $e = \frac{\rho * T}{216.7}$, and $r_p = \frac{p+e}{1013.25}$.

$$h_o = \frac{6.1 * A}{1 + 0.17 * r_p^{-1.1}} * (1 + t_1 + t_2 + t_3) \tag{3.22}$$

$$t_1 = \frac{5.1040}{(1 + 0.066 * r_p^{-2.3}} * exp[-(\frac{f - 59.7}{2.87 + 12.4 * exp(-7.9 * r_p)})^2] \tag{3.23}$$

$$t_2 = \sum_{i=1}^{7} \frac{c_i * exp(2.12 * r_p)}{(f - f_i)^2 + 0.025 * exp(2.2 * r_P)} \tag{3.24}$$

$$t_3 = \frac{0.0144 * f}{1 + 0.14 * r_p^{-2.6}} * \frac{15.02 * f^2 - 1353 * f + 5.333 * 10^4}{f^3 - 151.3 * f^2 + 9629 * f - 6803} \tag{3.25}$$

$$A = 0.7832 + 0.00709 * (T - 273.15) \tag{3.26}$$

$$h_w = A + B * \sum_{i=1}^{14} \frac{\alpha_i \sigma_w}{(f - f_i)^2 + b_i * \sigma_w} \tag{3.27}$$

$$A = 1.9298 - 0.04166(T - 273.15) + 0.0517 * \rho \tag{3.28}$$

$$B = 1.1674 - 0.00622(T - 273.15) + 0.0063 * \rho \tag{3.29}$$

$$\sigma_w = \frac{1.013}{1 + exp[-8.6 * (r_p - 0.57)]} \tag{3.30}$$

3.5.3 Total predicted attenuation by atmospheric gases

The attenuation can now be calculated for the zenith angle. The elevation angle β is accounted for by the cosecant law. This is only accurate for angle between 5°and 90°. The total predicted attenuation by atmospheric gases with inserted meteorological data gathered by BME is illustrated in Figure 3.4.

$$A = A_o + A_w = \gamma_o * h_o + \gamma_w * h_w \qquad [dB] \tag{3.31}$$

$$A = \frac{A_o + A_w}{sin(\beta)} \qquad [dB] \qquad 5^\circ \leq \beta \leq 90 \tag{3.32}$$

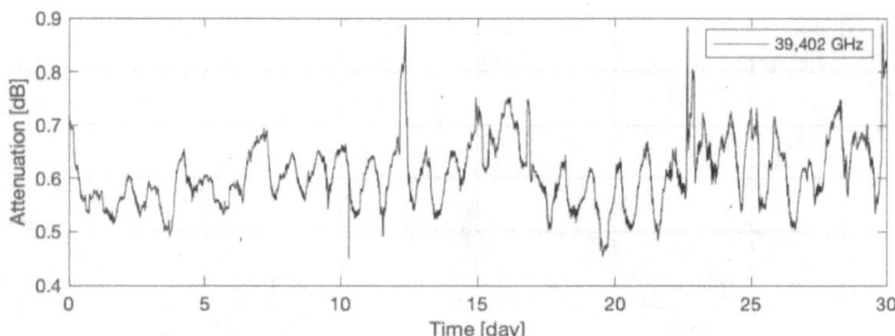

Figure 3.4: The predicted attenuation by atmospheric gases for Budapest location, April 2018. The gathered meteorological surface data from BME was applied on the method described in section 3.5.

3.6 Method of predicting attenuation by cloud and fog

The attenuation due to cloud and fog can be calculated by ITU recommendations [9]. The specific attenuation is given by the cloud liquid water specific attenuation coefficient K_l measured in units of $\frac{dB/km}{g/m^3}$ and the liquid water density in the cloud M in units of g/m^3. Two methods are given: one that requires the knowledge of the K_l and M over the vertical column and an approximation method with no required inputs. Equation (3.33) require the specific liquid water density in a specific cloud given by e.g. local radiosonde measurements or with a radiometer. The approximation method require no knowledge of the local parameters. The method calculates the attenuation with the total columnar content of cloud liquid water. Instead of the liquid water density M in equation (3.33) another parameter known as cloud liquid water (CLW) L in units of (kg/m^2) is used. It also set the temperature to a reduced value of $T = 273.15K$ since a cloud with temperatures below $0°C$ does not contribute to the attenuation to a large degree. Ice clouds formed over the 0° isotherm can thus be neglected. This reduced temperature parameter is written as L_{red}. This method only requires the frequency of the transmitted signal which makes it a useful tool for rough estimates of cloud attenuation when no locally gathered data is known.

$$\gamma_c(f,T) = K_l(f,T) * M \qquad [dB/km] \tag{3.33}$$

$$A_c = L_{red} * K_l(f, 273.15) \qquad [dB] \tag{3.34}$$

3.6.1 Digital maps

The CLW parameter L can be found for coordinates all over the surface of Earth given by digital maps that has been derived from ERA-40 data. These are meteorological data that has been gathered for decades since the 1950s. The digital maps provide data over the whole surface of Earth with a resolution of 1.125° that is from 0 to 360° in longitude and -90 to 90° in latitude. Any location on the surface is given by performing bi-linear interpolation from the four closest grid points to the location coordinates.
The CLW can be given over a full year or for the individual months. There are different probabilities based from the ERA-40 data of how likely it is to be reached. Monthly probabilities range from 1 to 99 % while yearly from 0.1 to 99 %.
If a low probability is chosen the likeliness of a large CLW value is increased. It also depends on geographical location as a more humid region, for example a rain forest region has a much larger probability to produce more CLW than a location in the polar regions.

3.6.2 Cloud liquid water specific attenuation coefficient, K_l

The CLW specific attenuation coefficient K_l is calculated by a mathematical model based on Rayleigh scattering which uses a double-Debye model for the dielectric permittivity $\epsilon(f)$ of water. Rayleigh approximation is valid for droplets with radius size of maximum 1/10 of the wavelength. Cloud and particles generally have a radius smaller than 0.01 centimeter which make the method valid for frequencies up to about 200 GHz[9].

$$K_l(f,T) = \frac{0.819 * f}{\epsilon''(1+\eta^2)} \qquad \left[\frac{dB/km}{g/m^3}\right]\Bigg(\tag{3.35}$$

$$\eta = \frac{2+\epsilon'}{\epsilon''} \tag{3.36}$$

ϵ' is the real part and ϵ'' the imaginary part of the dielectric permittivity for water. They are both frequency dependant and given by the liquid water temperature T, principal relaxation frequency f_p, and secondary principal relaxation frequency f_s. For the case of fog the surface temperature is used. For the approximation method, T is set to the fixed temperature of 273.15 K which only leaves one unknown parameter, the frequency f.

$$\epsilon'(f) = \frac{\epsilon_0 - \epsilon_1}{1 + (f/f_p)^2} + \frac{\epsilon_1 - \epsilon_2}{1 + (f/f_s)^2} + \epsilon_2 \tag{3.37}$$

$$\epsilon''(f) = \frac{f * (\epsilon_0 - \epsilon_1)}{f_p * (1 + (f/f_p)^2)} + \frac{f * (\epsilon_1 - \epsilon_2)}{f_s * (1 + (f/f_s)^2)} \tag{3.38}$$

$$\theta = 300/T \tag{3.39}$$

$$\epsilon_0 = 77.66 + 103.3 * (\theta - 1) \tag{3.40}$$

$$\epsilon_1 = 0.0671 * \epsilon_0 \tag{3.41}$$

$$\epsilon_2 = 3.52 \tag{3.42}$$

$$f_p = 20.20 - 146 * (\theta - 1) + 316 * (\theta - 1)^2 \tag{3.43}$$

$$f_s = 39.8 * f_p \tag{3.44}$$

3.6.3 Reduced cloud liquid water, L_{red}

The CLW temperature is assumed to be 273.15 K which make it possible to find K_l for any frequency between 1-200 GHz. L_{red} has to be chosen from any of the digital maps over a certain probability. A different probability can be found by performing bi-linear interpolation (see Equation (3.45)) of two probabilities. The algorithm for interpolation is by recommendation from ITU[14]. The elevation angle β is accounted for in the same way as for the gaseous attenuation by cosecant law. The elevation angle has to be in the range of $5° \leq \beta \leq 90$.

$$\begin{aligned} I(r,c) =& I(R,C)[(R+1-r)(C+1-c)] \\ &+ I(R+1,C)[(r-R)(C+1-c)] \\ &+ I(R,C+1)[(R+1-r)(c-C)] \\ &+ I(R+1,C+1)[(r-R)(c-C)] \end{aligned} \tag{3.45}$$

3.6.4 Total predicted attenuation by clouds

Equation (3.46) calculates the attenuation for a signal with frequency 39,402 GHz which is the same as for the received Alphasat signal. The location is based of coordinates for receiver station in Budapest as given by Table 1.1 and for an mean elevation angle of 35°. The L_{red} values are from the month of April given by recommendation from ITU[9]. The attenuation have no time dependence and a list for different probabilities for the month of April can be seen in Table 3.1 using the approximation method described above.

$$A_c = \frac{L_{red} * K_l(39.402, 273.15)}{sin(35^\circ)} \quad [dB] \tag{3.46}$$

Probability (%)	L (kg/m^2)	A_c (dB)
1	0.7515	1.4231
2	0.6606	1.2509
3	0.6111	1.1572
5	0.5248	0.9938
10	0.4152	0.7862
20	0.2580	0.4885
30	0.1510	0.2859
50	0.0272	0.0515

Table 3.1: Attenuation of cloud for different probabilities of CLW (L) for the month of April.

The following conclusions can be made from Table 3.1. For 1 % of the month of April an expected attenuation by clouds of 1.4231 dB will impair a signal with the same setup as the antennas operated by BME. 50% of the time there is an expected attenuation of only 0.0515 dB.

Chapter 4

Results and Discussion

Budapest, April 2018

The data analysed are from April 2018 which is the season of spring for Budapest. This can already set expectations for the reader of how the temperature profile will look. We can expect an increasing temperature over the course of the month with an oscillating diurnal cycle. The amount of humidity is not only dependant on the temperature, which will change the amount of saturation of water vapor in an air mass but also by larger effects which can not be predicted without long term observation of the area. That can include rain storms or other weather effects which are seemingly random in nature.

Surface meteorological data

Only the surface meteorological data is available which means there is no information of the vertical column of the atmosphere. The rain sensors can only sense rain that is precipitating directly above or in very close proximity. Since the antennas are pointing with an elevation angle towards the equator it can be expected a delay between the meteorological data and attenuation. The signal will travel through atmosphere that is not in a relationship with the surface instruments. A distant rain cloud that is approaching Budapest from the south would attenuate the signal before the rain sensor could indicate its presence. Since the cloud could approach from any direction an indication from wind parameters could give further information of expected time lag.

4.1 Distinction between rain attenuation and other effects

There is no method to determine the exact source of attenuation (i.e. cloud, gas, or scintillation) with only the surface meteorological data at hand. There is however possible to make distinction between rain and other atmospheric effects. This could be used to exclude attenuation by rain with high probability and find the effects of other phenomenon. The precipitation of rain can easily be picked up by surface parameters to pin-point the source. If a relationship between meteorological data and rain attenuation is established, the effect of other atmospheric effects can be studied in more detail. This section will start by characterise regions which are indicated by the rain sensors. It will further be analysed if the signal is showing patterns from other smaller attenuation effects. When the signal is characterised as other effects than rain it can be compared to the prediction methods provided by ITU.

4.2 Indication of rain by rain rate sensors

From the overview over the whole month of April 2018 seven different events can be identified by the rain rate sensors as seen in Figure 4.1b. Notice how the attenuation and rain rate are showing an increase at similar regions. These large attenuation levels can only be caused by rain as explained in earlier sections. The theoretical values of predicted gas and cloud attenuation (Figure 3.4 and Table 3.1) should not exceed a few dBs all together. Attenuation peaks in Figure 4.1a show regions with much larger dB reaching up to 60 dB.

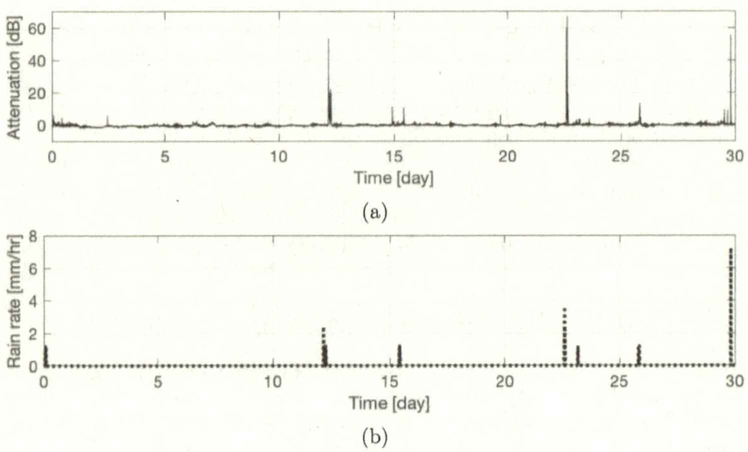

Figure 4.1: Comparison of attenuation and rain rate, April 2018.

4.3 The diurnal cycle of temperature and humidity

The diurnal cycle of humidity and temperature over April 2018 (Figure 4.2) show a typical spring month with an overall increasing trend for the temperature over the month. The humidity show a diurnal cycle with opposite trend compared to the temperature i.e. maximum humidity occur at minimum temperature. This show the strong relationship between the two parameters and how a lower temperature require less absolute humidity to reach saturation. There are a few days that deviate from the mean humidity with larger peaks. There are also small deviations between the individual days. These are interesting regions which indicate a disturbance from the diurnal cycle e.g. a rain event. The relationship between deviations in the surface meteorological data and attenuation is analysed to find if they influence each other.

4.3.1 Rain and surface humidity

When the rain rate (Figure 4.1b) is compared to the humidity (Figure 4.2b) a relationship can be expected. Peaks in relation to the x-axis at around 0, 12, 26, and 30 are all regions with increased rain rate and higher peaks of humidity. This suggest that the largest humidity levels are reached when rain occurs during a diurnal peak of humidity, i.e during late evening, midnight, or early mornings. If rain falls close to a minimum in the diurnal cycle around noon, a disturbance seems to occur and smaller peaks are seen e.g. at times 15,5 and 22,6 in Figure 4.2b. The wind velocity could also be a good indicator since large wind speed is typical for turbulent and stormy weather. This is however not investigated further. The auto-correlation can further indicate the cycles of the temperature and humidity. This will show similarities between the days.

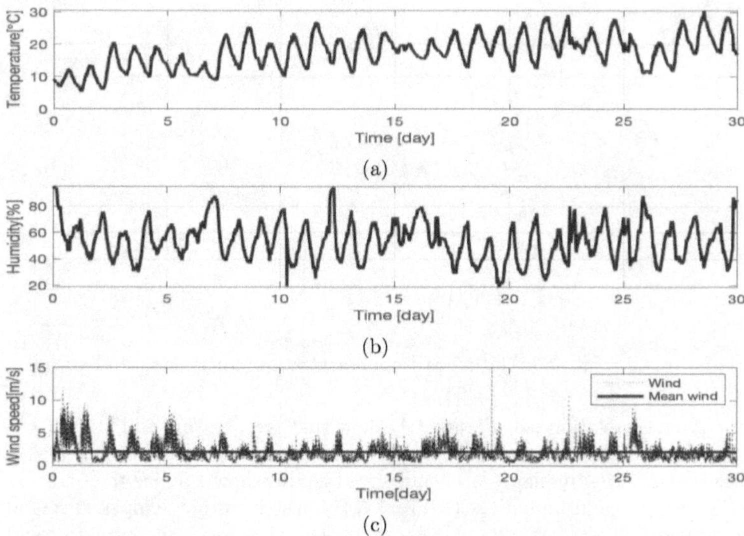

Figure 4.2: Comparison of temperature, relative humidity, and wind speed, April 2018.

4.3.2 Correlation between meteorological data

Notice in Figure 4.3 how the auto-correlation show maximum value of 1 at time lag 0 which is always the case while cross-correlation can find maximum values at different time lag. Also notice the difference in peaks in the auto-correlation function. This can be explained by how detailed a signal is. Since the humidity and temperature have a general smooth diurnal cycle the correlation does not decrease as much with a small shift as for the wind. The wind speed showing stronger fluctuations will show a larger decrease in the correlation function (see Figure 4.3c).

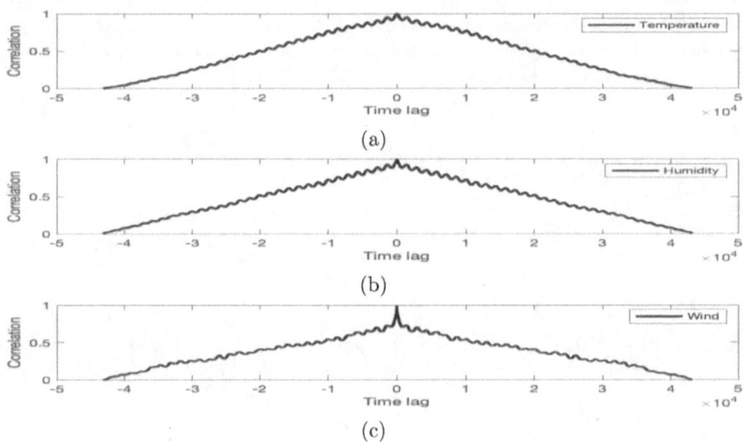

Figure 4.3: Auto-correlation of temperature, humidity, and wind speed.

Cross-correlation (see Figure 4.4) show that temperature and humidity is in a local minimum at zero time lag. The largest correlation is shown at a positive shifted time lag. The wind speed is also shown versus temperature (Figure 4.4b) and humidity (Figure 4.4c). This show that temperature and wind speed have maximum correlation at zero time lag while humidity and wind speed at negative shifted time lag. The triangle shape of the plots is due to the shifted signal slowly being zero-padded until completely shifted out by the vector length in either positive or negative direction.

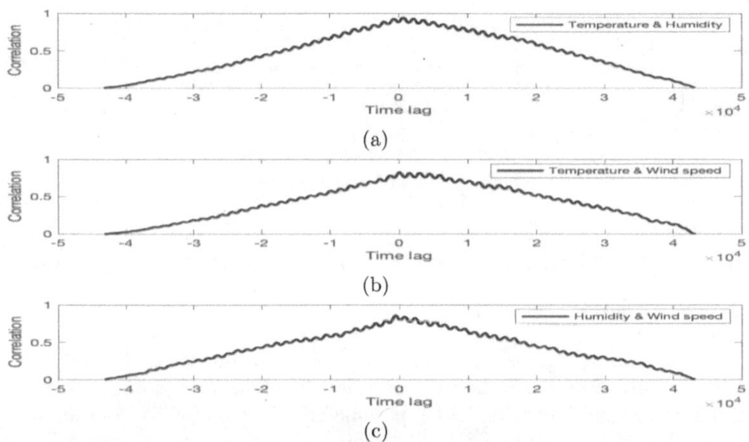

Figure 4.4: Cross-correlation of temperature/shifted humidity, temperature/shifted windspeed, humidity/shifted windspeed.

4.4 Finding attenuation events

There are many attenuation regions over the span of a month. A filtering script was run in Matlab to find region at certain threshold of attenuation. Four categories were created and are shown in the Table 4.1. The parenthesis show how many of these events are rain regions (with indicated rain rate in close proximity). The analysis starts with rain events to find a distinction from rain and other atmospheric effects. A study for regions with minimal attenuation (clear-sky) is also done to find natural characteristics of meteorological data. The remaining attenuation regions are then studied to see if we can establish a relationship for cloud and water vapor.

Category	Exceeded attenuation	Number of events
1	>10 dB	6(5)
2	10-5 dB	3(1)
3	5-2.5 dB	7(1)
4	2.5-1.5 dB	53

Table 4.1: Different categorises of attenuation events for April 2018.

4.5 Establishing rain attenuation characteristics

There are seven regions that were found with an increased rain rate (see Figure 4.1b). These regions are first studied to find a relationship between temperature and humidity. These parameters are expected to show an indication before the rain rate. The humidity can be expected to rise will the temperature drops. This could give indication of rain events where the rain clouds did not reach directly above to indicate rain sensor. Seven regions with rain is compared by its peak attenuation and peak rain rate (see Table 4.2).

Time (day)	Peak Att. (dB)	Peak Rain (mm/hr)
0-0,4	06,29	1,2
12,1-12,3	53,59	2,4
15,3-15,5	10,75	1,2
22.59-22.64	67,45	3,6
23-23,25	04,09	1,2
25,6-26	13,70	1,2
29,8-29.85	55,72	7,2

Table 4.2: Regions with indicated rain rate above 0 mm/hr, April 2018

Three of the regions show a very large attenuation above 50 dB. They also show the largest rain rate, however the largest rain rate does not correspond to the largest attenuation. The seven regions are divided into two figures (see Figure 4.5 and Figure 4.6). The left y-axis show the attenuation in dB while the right show the rain rate in mm/hr. Keep in mind that the axis are not fixed for the plots and adjusted for each event to show more detail.

The relationship between rain rate and attenuation is now clearly visible. Keep in mind that a shift by 0,04 on the x-axis correspond closely to one hour in real time. There are both rain rate indicated over short periods of time (see Figure 4.5) and long periods of time (see Figure 4.6. The short rain duration seem to correspond to lower attenuation levels. The long rain duration correspond to larger attenuation levels, an expected and logical behavior.

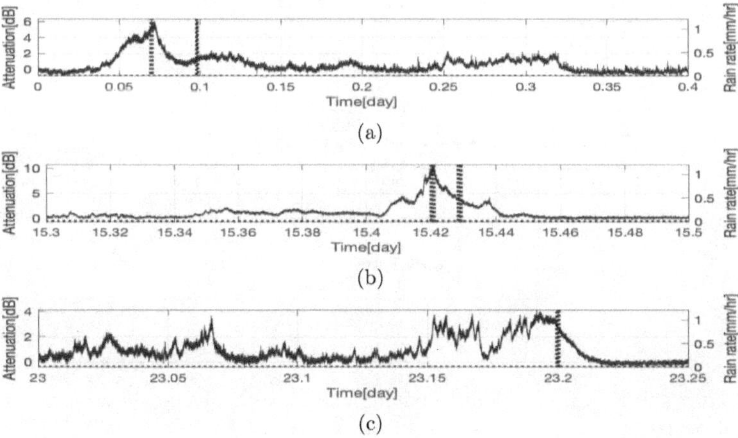

Figure 4.5: Three rain regions listed in Table 4.2 with short duration, April 2018.

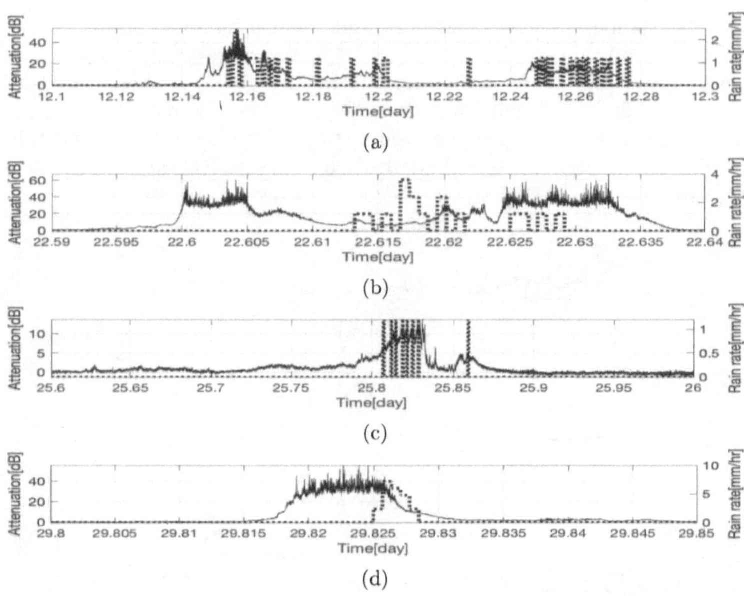

Figure 4.6: Four rain regions listed in Table 4.2 with a longer duration, April 2018.

Difference between zenith path and signal path with 35 ° elevation angle

Even though the rain rate give a definite indication of the source of rain attenuation it seems to become less accurate at lower attenuation levels. It is also not guaranteed that rain will be captured by the rain sensors. Since they are located at the antennas, only rain falling directly above will be indicated. The antennas in Budapest are elevated with a mean elevation angle 35° which means that the path of the signal does not correspond to the path of the rain rate sensors. If a rain cloud would only pass through the signal path and not above the antennas its presence would remain unknown by the sensors.

4.5.1 Comparison: attenuation/temperature, attenuation/humidity

The humidity and temperature can also be strong indicators of rain events. The time regions in Table 4.2 are increased are studied since a larger interval illustrate disturbance from the diurnal cycle more clearly. The rain regions in Table 4.2 were extended by a value of 0,5 (12 hours) in both negative and positive direction increasing the interval by one day (see Figure 4.7-Figure 4.12). This will show if any disturbance from the natural cycle is occurring during these regions. One of the regions (see Figure 4.8) include two regions as they occurred within half a day of each other. This figures thus almost stretch over two days between time 22-23,8 on the x-axis.

Closer look at the three largest rain events exceeding 50 dB

The three largest attenuation regions shown in Figure 4.7-Figure 4.9 indicate a strong disturbance from the diurnal cycle. The temperature drops by several ° C and humidity increase by tens of percents. The small rain event included in Figure 4.8 at around time 23,2 seem to show a similar behavior but to much a lesser extent.

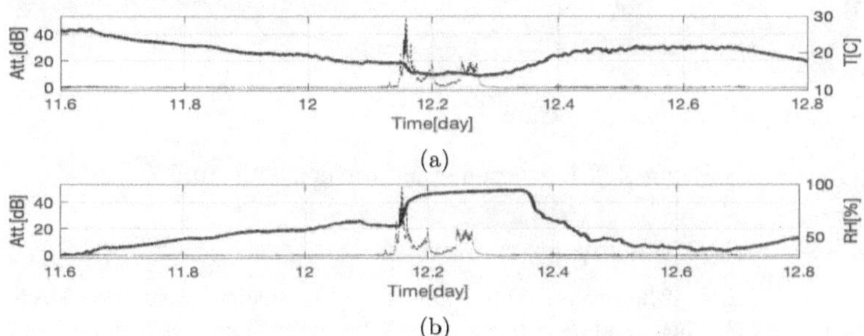

(a)

(b)

Figure 4.7: Large rain event, morning 13th April.

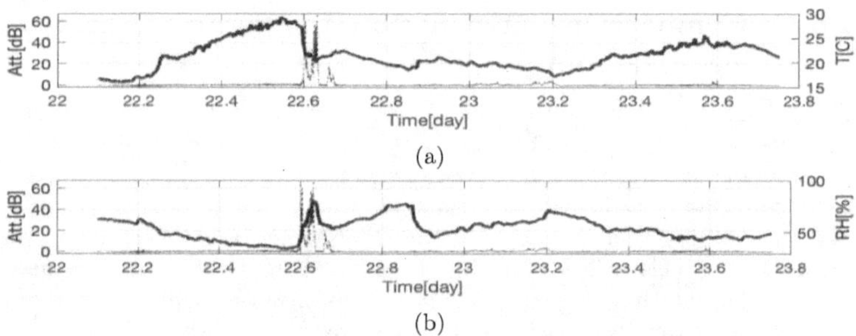

Figure 4.8: Large rain event, noon 23rd April.

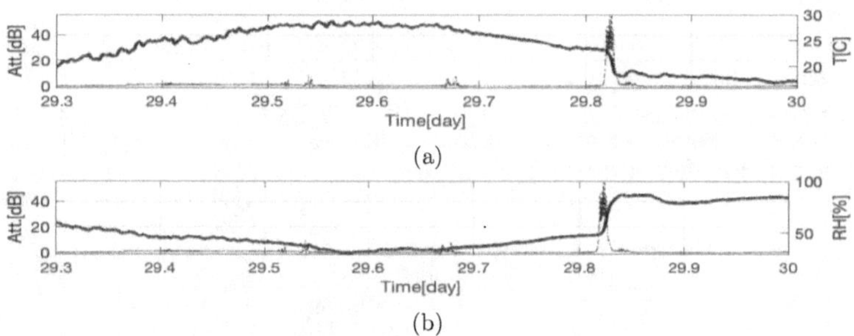

Figure 4.9: Large rain event, evening 30th April.

Closer look at the smaller rain events

There are three other events from Table 4.2 with smaller attenuation levels below 14 dB. They all show a rain rate of 1,2 mm/hr. The first region shown in Figure 4.10 show a large initial humidity close to 100%. This suggest an already ongoing event from previous month of March. The surface humidity would not normally reach these levels without precipitation of rain. This region is thus not a good example since it does not portray the behavior of a full scale rain event but the aftermath.

Figure 4.11 does not only show the rain event around 15,4 on the time line but also another similar attenuation event half a day earlier. This could show another rain event that was not captured by the rain rate sensors. The similar exponential attenuation shape and behavior of meteorological data suggest this.

Figure 4.12 does not show a clear disturbance in the diurnal cycle although it reaches one of the largest humidity levels (> 80%) during the month (see Figure 4.2b). The rain event occurs during the evening and so will the humidity increase naturally by the diurnal cycle. The duration of the rain event is over one hour as indicated by Figure 4.6c. The constant rain rate of 1,2 mm/hr and long period suggest stratiform rain i.e. large spatially covering rain clouds. This could also explain the smoother increase of humidity and temperature with no extreme change in meteorological data.

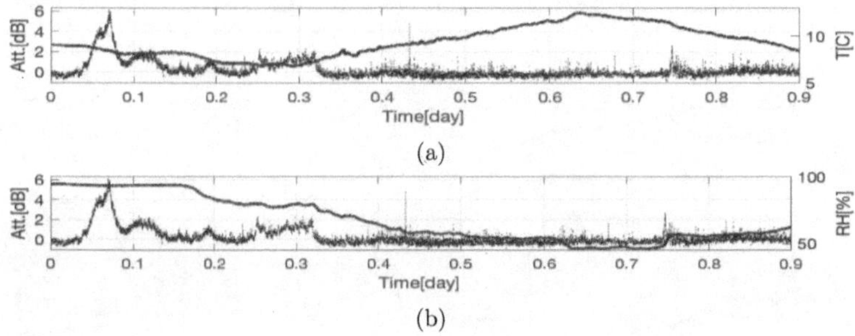

Figure 4.10: Small rain event, morning 1st April. Large humidity close to 100% already visible at the start suggesting ongoing rain event.

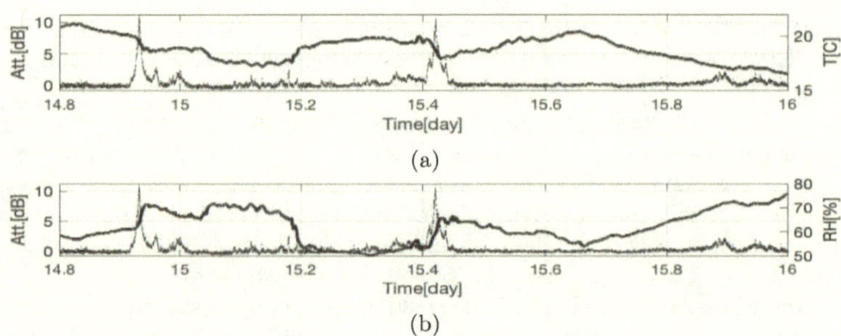

(a)

(b)

Figure 4.11: Two smaller rain events within half a day of each other, midnight 15th and noon 16th April.

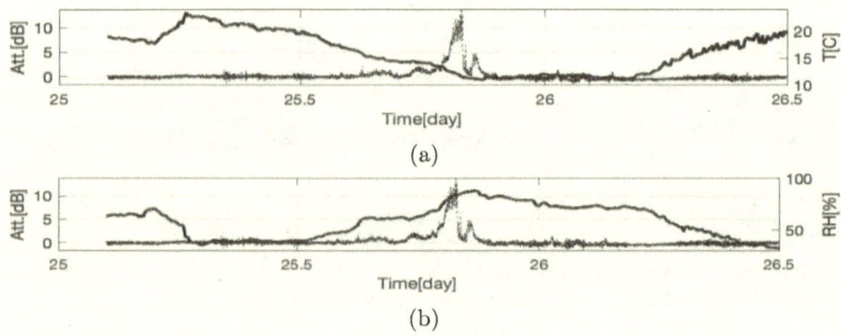

(a)

(b)

Figure 4.12: Smaller rain event with smooth change in meteorological data, evening 26th April.

4.5.2 Cross-correlation: temperature/shifted humidity

The further establish a relationship between humidity and temperature during rain events the cross-correlation is calculated. The humidity is shifted over the temperature. The shorter regions defined in Table 4.2 are once again used since the behavior close the to events are of interest. The seven regions are shown in Figure 4.13-Figure 4.19. All but one of the regions show a distinct negative correlation with decreasing temperature and increasing humidity. That indicate that they have strong opposite behavior i.e. when temperature drops humidity increase and vice versa. This strong relationship is one of the characteristic of rain events.

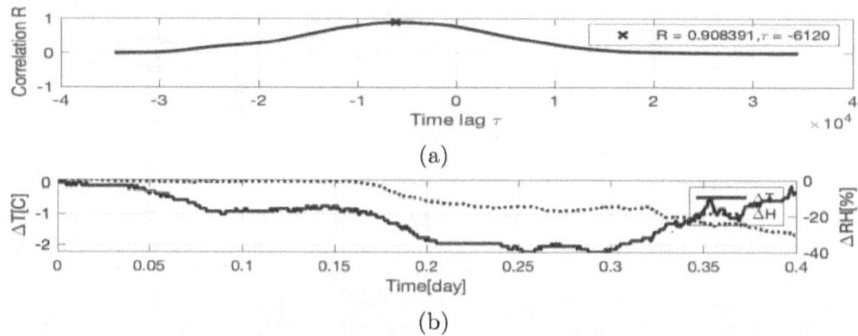

Figure 4.13: Positive correlation due to the aftermath of an already ongoing rain event from March. As the rain stops pouring the surface humidity slowly drops back to normal levels during morning on 1st April. Negative time lag show a larger correlation with shifted humidity as the temperature drops before the humidity.

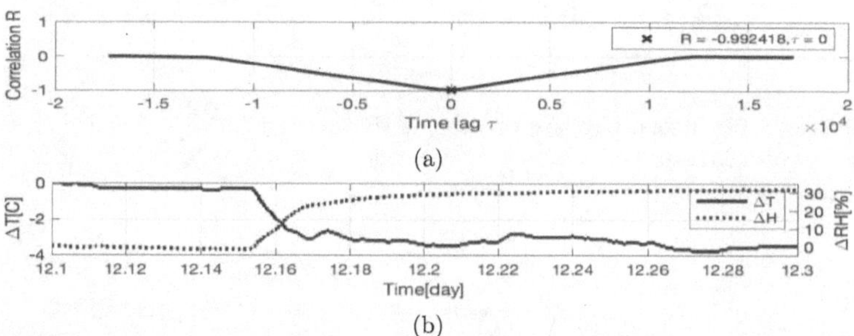

Figure 4.14: Strong negative correlation (R=-0,99) with zero time lag (τ = 0), midnight 12th April.

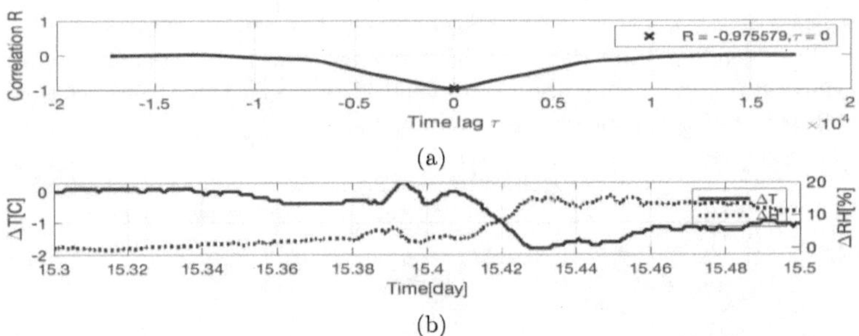

(a)

(b)

Figure 4.15: Strong negative correlation (R=-0,975) with zero time lag ($\tau =$ 0), noon 16th April.

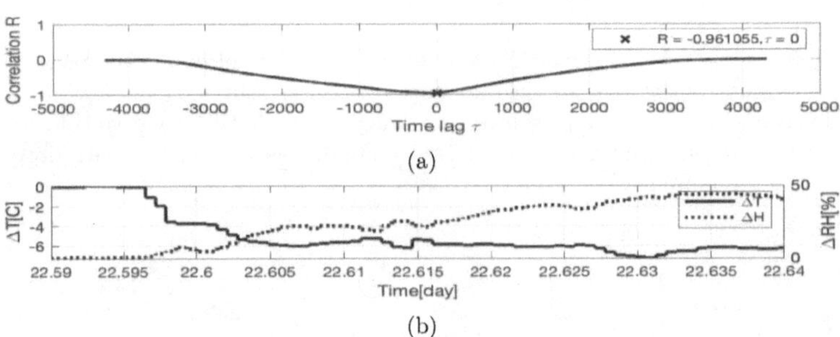

(a)

(b)

Figure 4.16: Strong negative correlation (R=-0,96) with zero time lag ($\tau =$ 0), noon 23rd April.

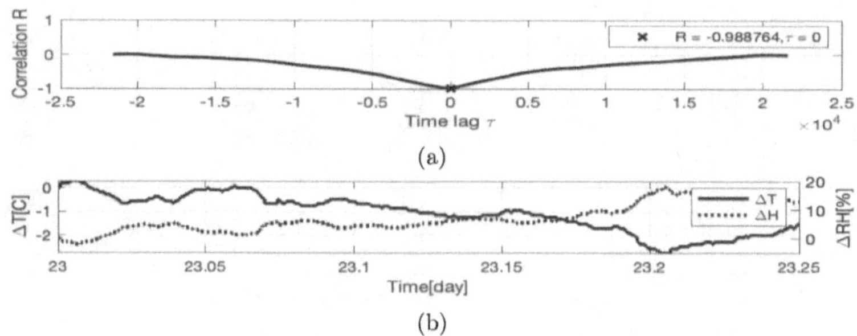

Figure 4.17: Strong negative correlation (R=-0,988) with zero time lag ($\tau = 0$), early morning 24th April. This is mostly due to the diurnal cycle with small rain event at 23,2 showing small disturbance.

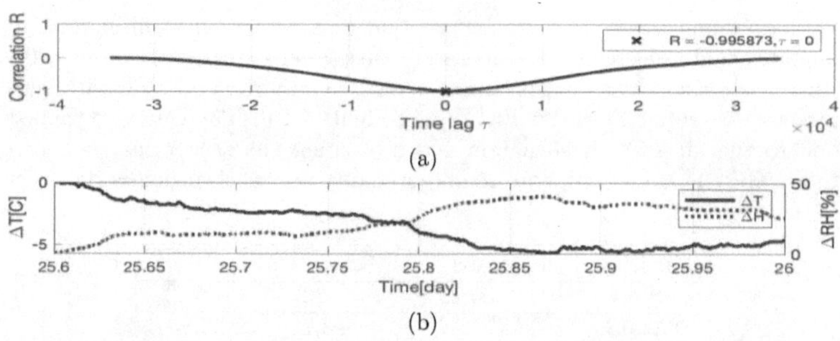

Figure 4.18: Strong negative correlation (R=-0,995) with zero time lag ($\tau = 0$), evening 26th April.

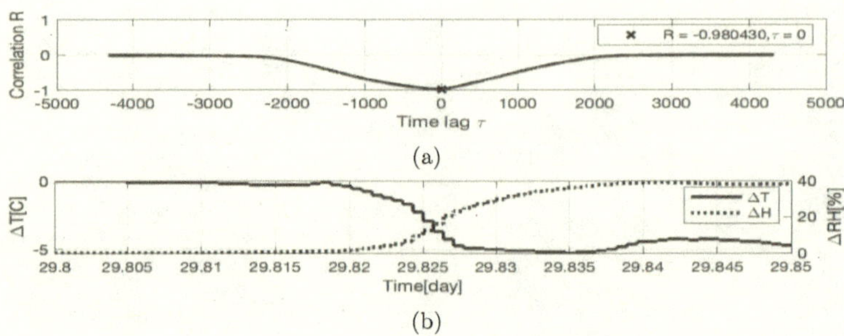

(a)

(b)

Figure 4.19: Strong negative correlation (R=-0,98) with zero time lag ($\tau = 0$), evening 30th April.

Increase in humidity and decrease in temperature

We have found a general characteristic that a rain event show an increase in humidity and a decrease in temperature with good synchronisation ($\tau = 0$). All regions excluding the first event, show maximum correlation with zero time lag (see subfigures (a) in Figure 4.14-Figure 4.19). This can be explained due to the air mass from a rain event reaching the sensors at the same time. The change of temperature and humidity over each region is shown in Table 4.3.

Time (day)	Peak Att. [dB]	ΔT [°C]	ΔRH[%]
0-0,4	06,29	-2,3	-30,23
12,1-12,3	53,59	-3,75	33,85
15,3-15,5	10,75	-2,08	16,62
22.59-22.64	67,45	-7,35	43,96
23-23,25	04,09	-3,02	20,23
25,6-26	13,70	-5,70	40,25
29,8-29.85	55,72	-5,31	39,32

Table 4.3: The change of temperature and humidity over the rain regions. A negative value indicate a decrease while positive an increase.

4.5.3 Cross-correlation: attenuation/shifted temperature, attenuation/shifted humidity

The most interesting aspect is if there is a relationship between attenuation and meteorological data. We know from previous sections that there is a disturbance in meteorological data in relation to large attenuation events but not how well it correlates with the attenuation. Both the temperature and humidity is shifted over the attenuation on the same seven rain regions as shown in Table 4.2. The correlation function over the seven regions show similar behavior excluding the first region (see Figure 4.20). Full line represent attenuation compared with shifted temperature while the dotted line represent attenuation compared with shifted humidity. A large positive correlation for the dotted line indicate that both humidity and attenuation increase over that region. A large negative correlation for the full line indicate that the temperature decrease while attenuation increase. The function show a mirror-like behavior which is due to the good synchronisation of temperature and humidity shown in earlier section. This will conclude the characterisation of the indicated rain regions from the rain rate sensor. Other regions with similar behavior can be assumed to be from rain. The correlation function is also shown together with attenuation and temperature/humidity over the regions in Appendix A section A.1.

Characteristics of rain

- Strong negative correlation between temperature and humidity.
- Strong positive correlation between attenuation and humidity with maximum at slight negative time lag.
- Strong negative correlation between attenuation and temperature with maximum at slight negative time lag.
- Plot with both correlation function create an "eye" due to the mirror-like behavior of meteorological data.

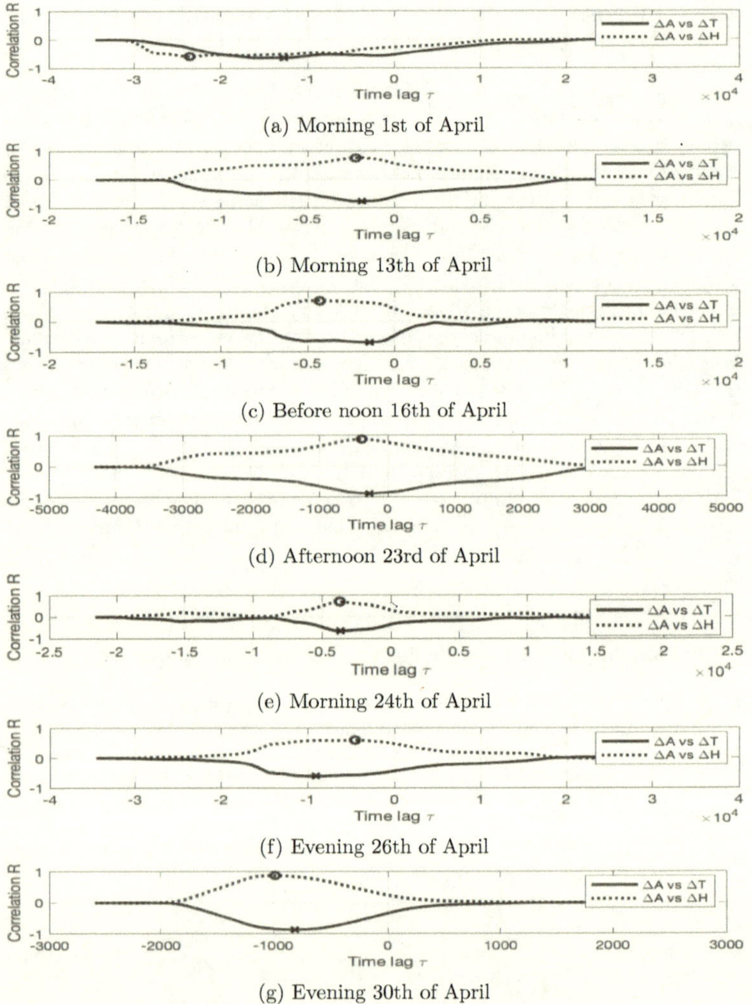

(a) Morning 1st of April

(b) Morning 13th of April

(c) Before noon 16th of April

(d) Afternoon 23rd of April

(e) Morning 24th of April

(f) Evening 26th of April

(g) Evening 30th of April

Figure 4.20: The correlation functions for the seven rain regions in chronological order as shown in Table 4.2.

4.6 Establishing characteristics for clear-sky regions

It is of interest to see how the meteorological data behaves during low attenuation, i.e during clear-sky conditions. This identifies naturally behavior in humidity and temperature. Ten regions were manually chosen were no visible attenuation event occurred (see Table 4.4). Eight of the regions occur during evening/midnight/early morning which is peak for humidity and minimum for temperature in the diurnal cycle. This behavior is also represented in Table 4.4 by the negative temperature values and positive humidity values. Two of the regions occur during noon which is maximum for temperature and minimum for humidity in the diurnal cycle. This behavior is also indicated in the table with positive temperature values and negative humidity values.

Time (day)	ΔT [°C]	ΔRH[%]
1,8-2,2	-4,06	22
2,9-3,1	-2,05	3,9
3,4-3,6	2,64	-10,7
12,8-13,2	-6.83	27,7
17,8-18,2	-7,43	27
18,4-18,7	5,91	-7,43
19,8-20,2	-7,69	31,7
20,8-21	-5,84	19,5
23,8-24,2	-6,87	25,5
26,7-27	-6,11	22,6

Table 4.4: The 10 clear-sky level regions and the corresponding change of meteorological data over those regions.

4.6.1 Comparison: attenuation/temperature, attenuation/humidity

Four regions were chosen to represent the general behavior due to their similarity. Figures over all other regions can be found in Appendix B section B.1. Figure 4.21 show a region during midnight between 3rd and 4th of April. All of the midnight regions show a similar behavior of decreasing temperature and increasing humidity. Small fluctuations can occur in both parameters but are restricted to a few percentages for humidity and generally below 1° C. Figure 4.22 and Figure 4.23 show both regions over noon. Figure 4.24 is a region during evening which also show clear pattern from the tracking system. This system operates periodically and keeps the pointing error between the Alphasat satellite and antenna below +/- 0,5 dB.

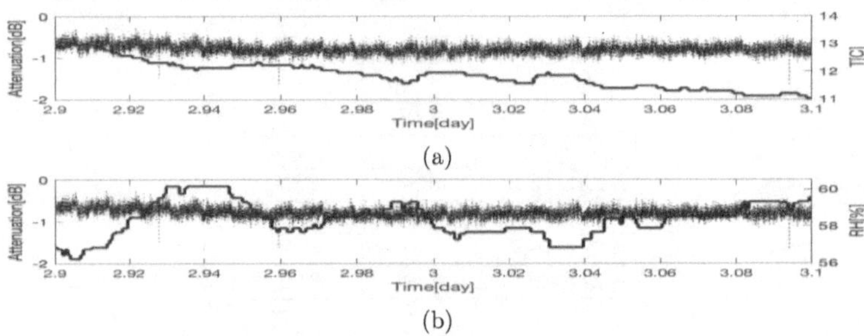

(a)

(b)

Figure 4.21: Small fluctuations are shown in the humidity while temperature decrease fairly steady, midnight 3rd of April. .

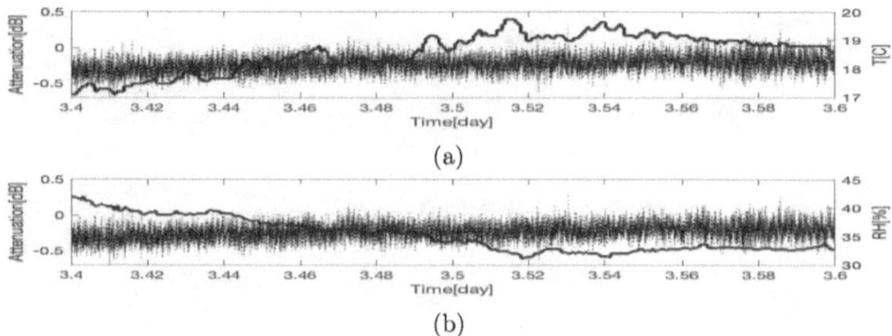

Figure 4.22: Small fluctuations in the temperature while humidity show a steady curve, noon 4th of April.

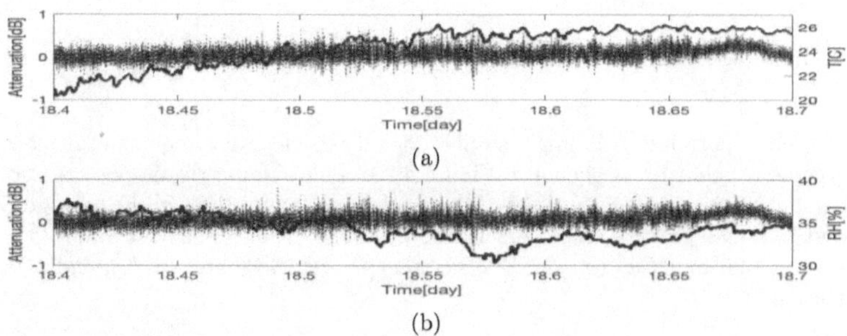

Figure 4.23: Small fluctuations are shown on both temperature and humidity, noon 19th of April.

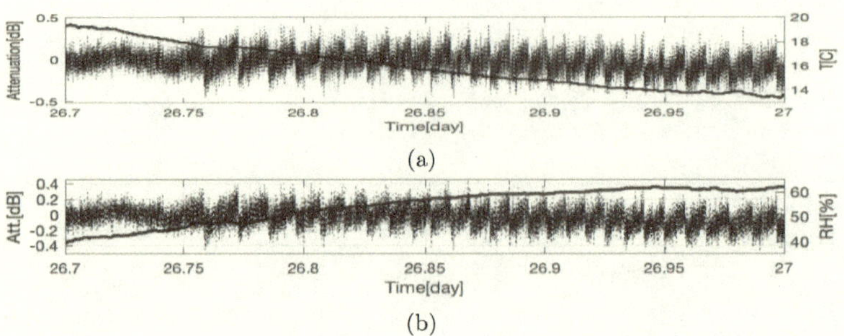

Figure 4.24: Steady decline of temperature and incline of humidity, evening 27th of April. A clear pattern of how the tracking system operates periodically and adjusting direction if pointing error reaches +/- 0,5 dB.

4.6.2 Cross-correlation: temperature/shifted humidity

The cross-correlation is fairly similar for all ten clear-sky regions and have a general behavior as shown in Figure 4.25. This symmetric shape was also similar for the rain events. Even if there are small fluctuations in the humidity and temperature the cross correlation will show a high negative value (see Figure 4.26 and Figure 4.27). Other regions can be found in Appendix B section B.2.

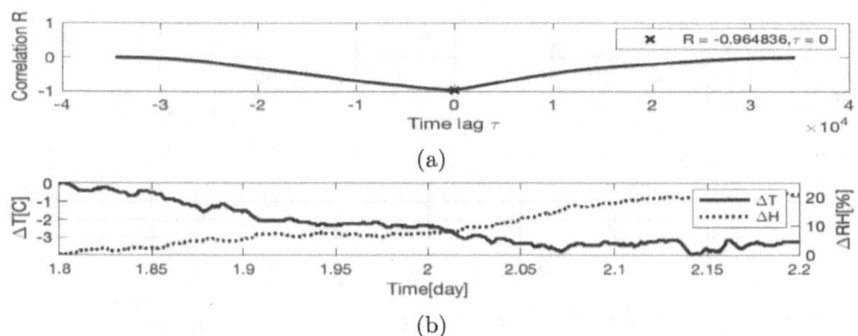

(a)

(b)

Figure 4.25: General characteristic of the midnight regions. Decrease in temperature and increase in humidity, midnight 2nd of April.

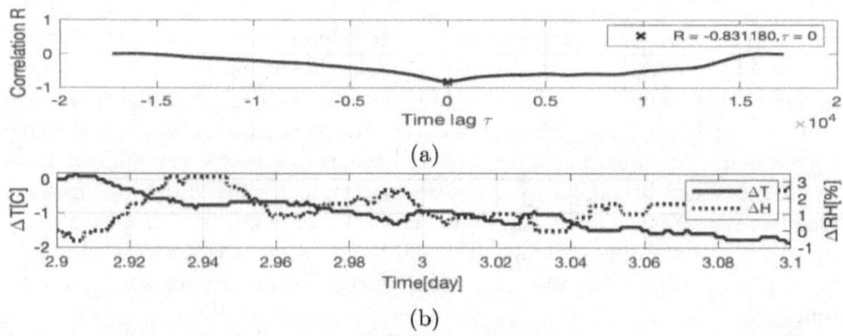

(a)

(b)

Figure 4.26: Moderate fluctuations which yields lower correlation $R = -0,83$ but in general similar trend decrease of temperature and increase of humidity, midnight 3rd of April.

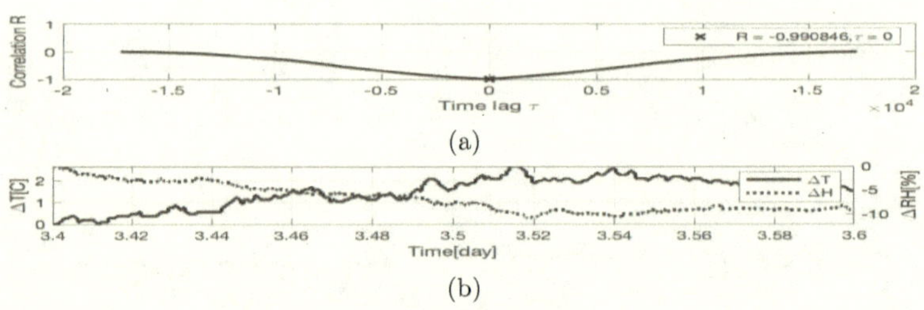

(a)

(b)

Figure 4.27: General characteristic of noon regions. Increase in temperature and decrease in humidity, noon 4th of April.

4.6.3 Cross-correlation: attenuation/shifted temperature, attenuation/shifted humidity

Three figures are shown to represent the ten different cross-correlation functions. Other regions are found in Appendix B section B.3. Due to eight of the regions being around midnight, the general correlation function looks similar to Figure 4.28. The attenuation and temperature show a positive correlation. This might not be expected but is due to the attenuation fluctuating at 0 dB. It slightly decrease over the region which provides a general negative dB value. Attenuation and temperature being both negative sums up to a positive value. The overall trend is thus a positive correlation attenuation/temperature function and negative correlation attenuation/humidity function.

The two noon regions are shown in Figure 4.29 and Figure 4.30. Notice how both regions show similar behavior for the meteorological data but the correlation function does not. This is due to the attenuation being mostly positive in Figure 4.29 while it is mostly negative in Figure 4.30. The correlation is thus some what unreliable measurement as it is hard to distinguish visually the behavior of attenuation i.e if it is positive or negative.

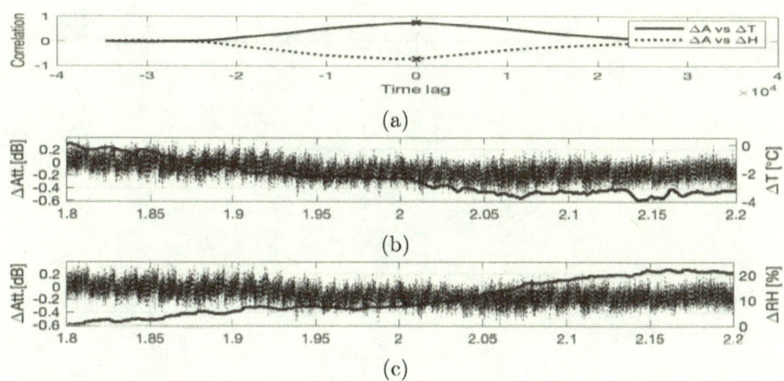

Figure 4.28: Symmetric correlation function and attenuation versus temperature/humidity over midnight on the 2nd of April.

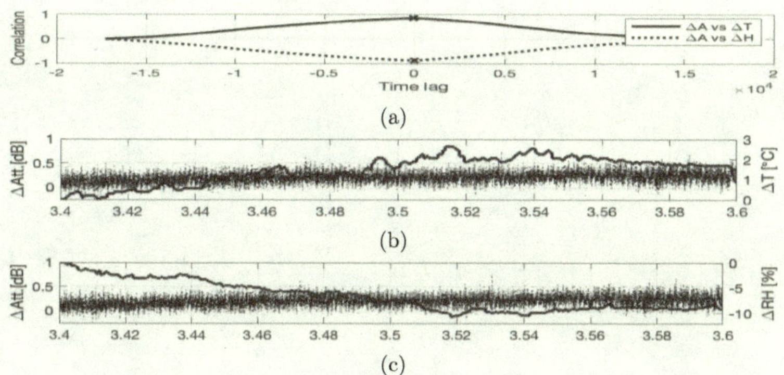

Figure 4.29: Symmetric correlation function and attenuation versus temperature/humidity over noon on the 4th of April.

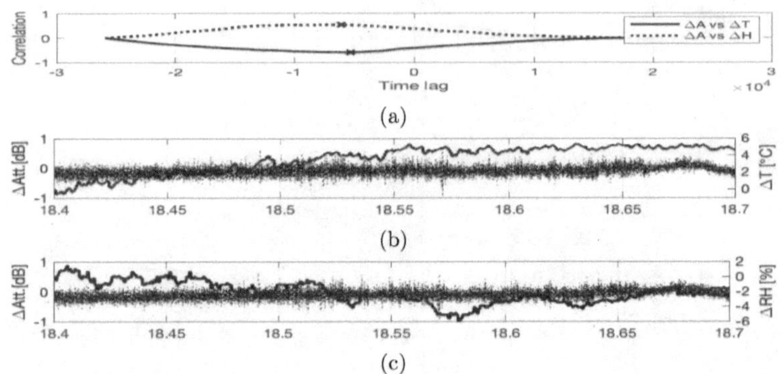

Figure 4.30: Symmetric correlation function and attenuation versus temperature/humidity over noon on the 19th of April.

9 of the 10 clear-sky regions show a similar reversed correlation function as for the rain event. Figure 4.30 show similar correlation function as for rain but the function also show a much smoother decrease over time lag τ.

That is due to the stable increase/decrease which is not seen for rain events. The correlation function for temperature and humidity show the strong relationship of the diurnal cycle. We see that a high and broad correlation function can be found. This broader correlation function imply the smooth diurnal cycle. There can be found regions where the two meteorological data change independently. A characteristic that was less common during rain event as they showed strong dependency. However, the correlation function portray the strong general dependency of the diurnal cycle and these small individual changes are not visible.

Characteristics of clear-sky

- Strong negative correlation between humidity and temperature due to the diurnal cycle.
- An independently behavior of temperature and humidity can be found for some regions.
- Independent fluctuations are not visible on the temperature/humidity correlation function if the region is too large due to diurnal cycle.

4.7 Establishing cloud and water vapor characteristics

The first non-rain attenuation events that are investigated are for higher attenuation levels without indication from rain rate sensors (see Table 4.1). There were 9 events above 2,5 dB that had no indication from rate rain sensors as shown in Table 4.5. We look for regions with other characteristics than was found for rain. These are regions that may show attenuation events from cloud and water vapor.

Time (day)	Temperature ΔT (°C)	Humidity $\Delta RH(\%)$
0,42-0,45	0,56	-4,02
0,73-0,79	-1,30	10,22
6,15-6,55	2,98	15,97
14,9-15,03	-1.61	11,15
15,15-15,21	1,70	-15,89
23,57-23,61	-2,10	4,54
28,76-28,79	-0,67	-2,06
29,5-29,57	1,54	-9,91
29,66-29,69	-1,44	2,89

Table 4.5: Regions with no indication of rain rate with attenuation above 2,5 dB.

4.7.1 Comparison: attenuation/temperature, attenuation/humidity

Four of the nine regions from Table 4.5 are presented in figures below. All of the other regions are found in Appendix C section C.1. There are no distinct general behavior over the regions. The humidity and temperature can now change separately as indicated by Figure 4.31, Figure 4.33, Figure 4.34. The humidity is seen as increasing to a larger degree than the temperature which is showing a steady decline throughout the region. Figure 4.32 show typical rain characteristics with a mirror-like response in both temperature and humidity. Figure 4.33 and Figure 4.34 show fluctuating temperature while the humidity show a linear decrease/increase.

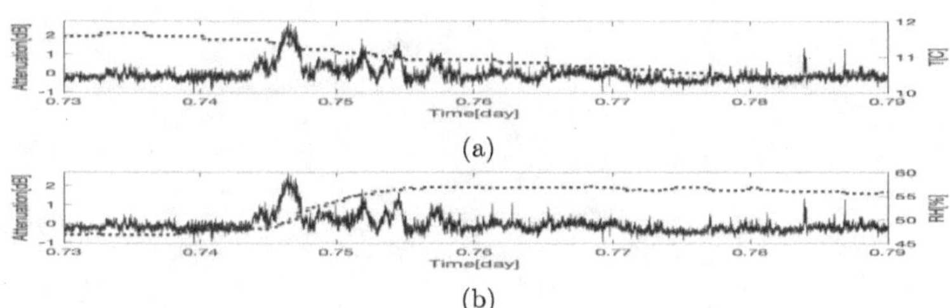

Figure 4.31: 2 dB attenuation event in the evening on 1st of April.

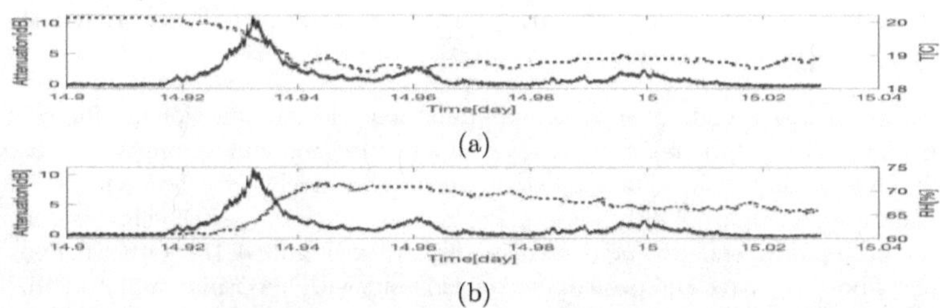

Figure 4.32: 10 dB attenuation event before midnight on 15th of April.

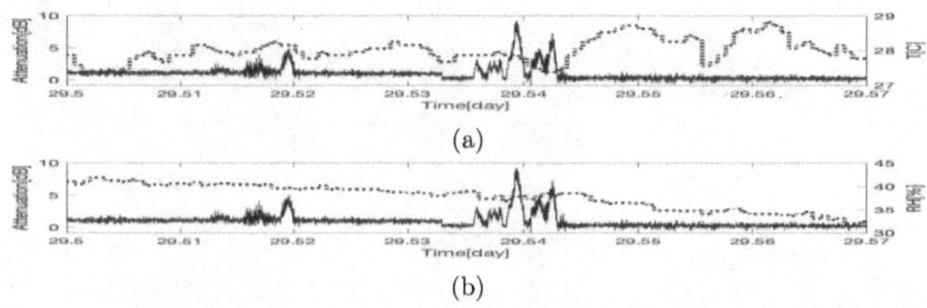

Figure 4.33: 9 dB attenuation event after noon on the 30th of April.

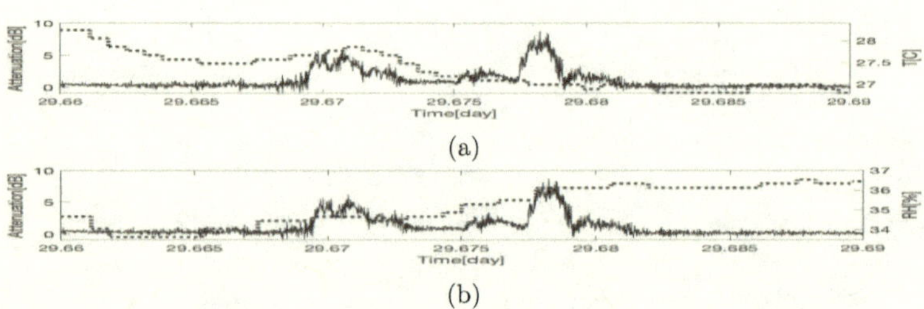

Figure 4.34: 6 dB attenuation event in the evening on the 30th of April.

4.7.2 Cross-correlation: temperature/shifted humidity

There are four regions that show similar correlation function as for rain events. Three of the regions show a temperature drop and humidity increase while following each other with strong negative correlation (see Figure 4.35, Figure 4.36, Figure 4.38). One of the regions show opposite behavior with increasing temperature and decreasing humidity (Figure 4.37). Other regions show both negative and positive correlation but with lower maximum R value and at different time lag τ (see Figure 4.38-Figure 4.43). These regions show lower correlation ($|R| < 0,9$) and/or asymmetric correlation function. These are the best candidates for being cloud and water vapor attenuation.

Regions that show similarity to rain characteristics

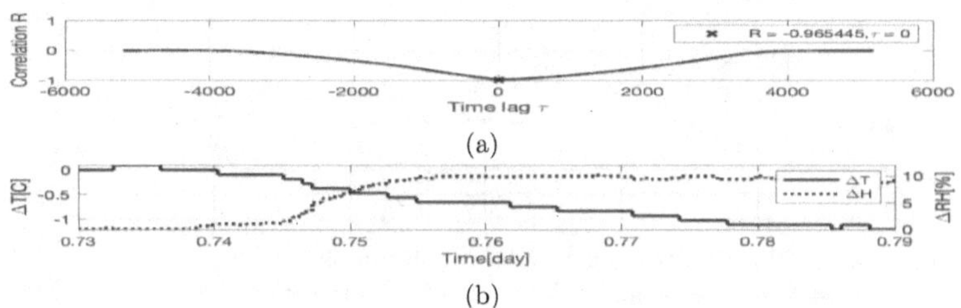

Figure 4.35: Strong negative correlation (R=-0,965) at zero time lag ($\tau = 0$), evening on 1st of April. Symmetric correlation function.

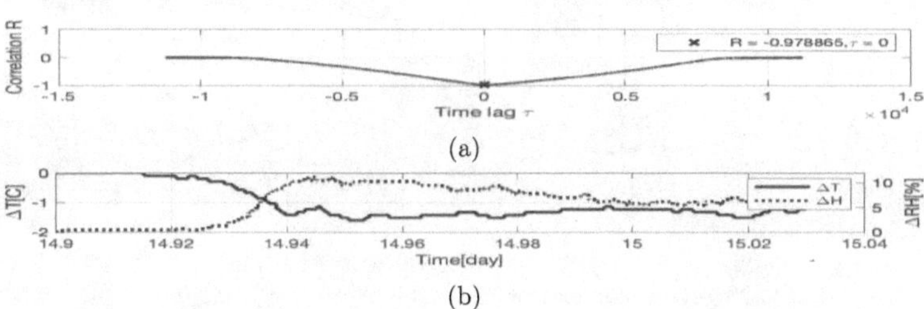

Figure 4.36: Strong negative correlation (R=-0,978) at zero time lag ($\tau = 0$), before midnight 15th of April. Symmetric correlation function.

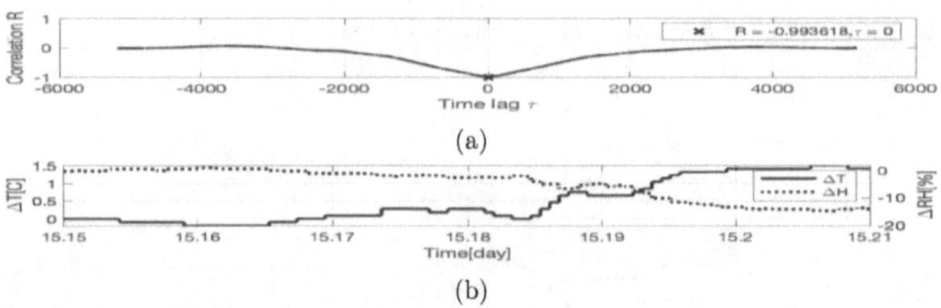

Figure 4.37: Strong negative correlation (R=-0,994) at zero time lag ($\tau = 0$), early morning 16th of April. Symmetric correlation function but with increasing temperature and decreasing humidity.

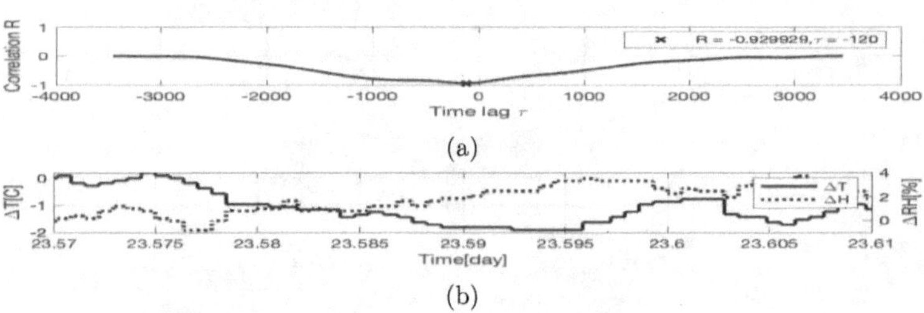

Figure 4.38: Negative correlation (R=-0,929) at negative time lag ($\tau = -120$), afternoon 24th of April. Symmetric correlation function.

Regions that differ from rain characteristics

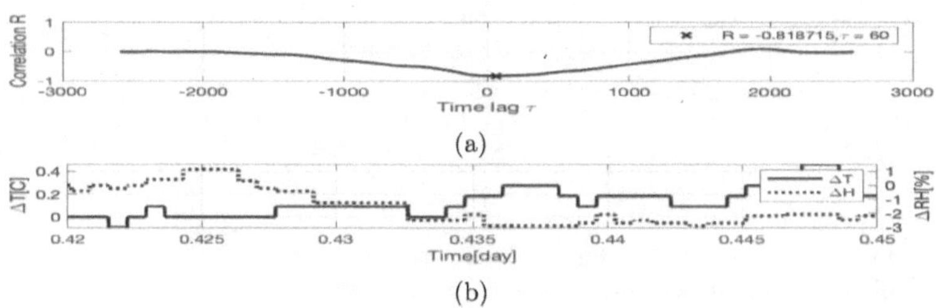

Figure 4.39: Negative correlation (R=-0,82) at positive time lag (τ = 60), before noon on 1st of April. Symmetric but not high enough correlation.

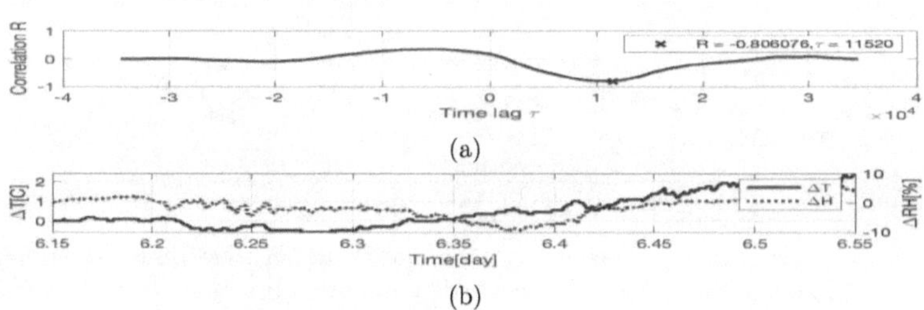

Figure 4.40: Negative correlation (R=-0,806) at positive time lag (τ = 11520), early morning on 7th of April. Asymmetric correlation function.

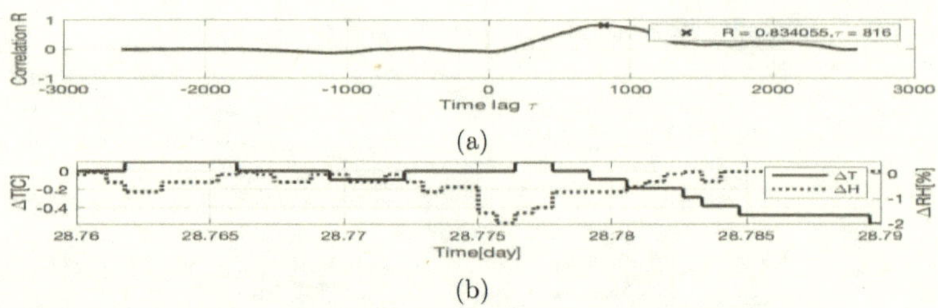

Figure 4.41: Positive correlation (R=0,834) at positive time lag ($\tau = 816$), evening 29th of April. Asymmetric correlation function.

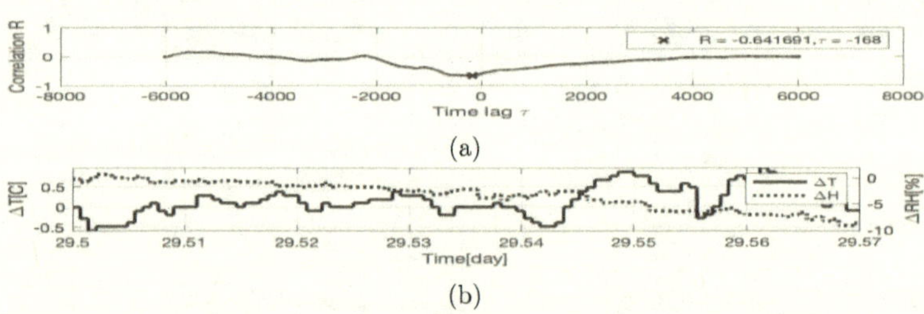

Figure 4.42: Negative correlation (R=-0,642) at negative time lag ($\tau = -0,64$), noon 30th of April. Asymmetric correlation function.

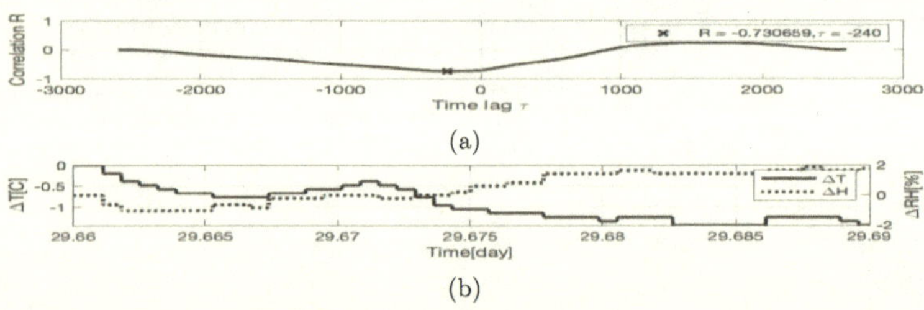

Figure 4.43: Negative correlation (R=-0,731) at negative time lag ($\tau = -240$), afternoon 30th of April. Asymmetric correlation function.

4.7.3 Cross-correlation: attenuation/shifted temperature, attenuation/shifted humidity

The four regions in previous section that had symmetric/mirror-like response in temperature/humidity similar to rain are seen in Figure 4.44-Figure 4.47.

Figure 4.44 show similarity in correlation function (e.g. a lower correlation version of Figure 4.20g) to a rain event but with lower peak correlation. This depends on the fast rise of attenuation while especially temperature have a slower decline. It is only an attenuation level of 2 dB which means that even if the source is rain, the amount of rain is small and might not show indication in rain sensors. Similar levels of decrease in temperature and increase in humidity was seen in clear-sky regions during evening e.g. Figure 4.24 which makes the source indistinguishable from the diurnal cycle. However, the increase in humidity close to the attenuation could suggest a small rain event.

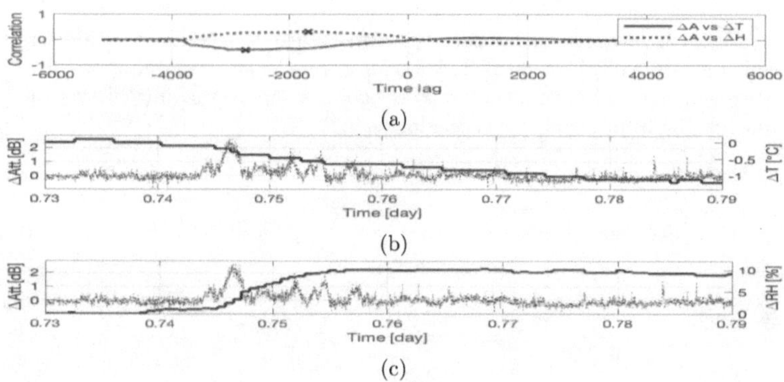

Figure 4.44: Symmetric correlation function shape but with low values ($|R| < 0,5$). The humidity show an increase close to the attenuation event while temperature steady drops over the region, evening 1st of April.

Figure 4.45 show clear similarities with rain events and were already discussed as a potential rain event in previous sections. All the parameters except an indication by rain rate sensors are similar to the rain events. It can thus be assumed to be attenuation by rain.

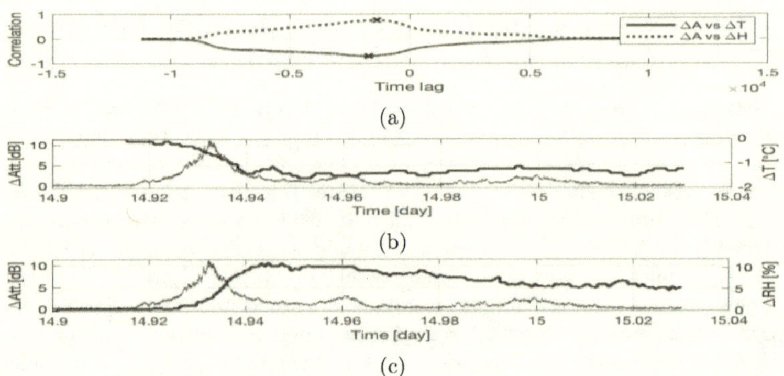

Figure 4.45: High correlation for attenuation with both temperature and humidity. A clear rain characteristic behavior, before midnight 15th of April.

Figure 4.46 show an increase of temperature and decrease of humidity which is not typical for rain. This region is on the early morning of 16th of April which would naturally show this behavior due to the diurnal cycle and can thus be from cloud/atmospheric gases.

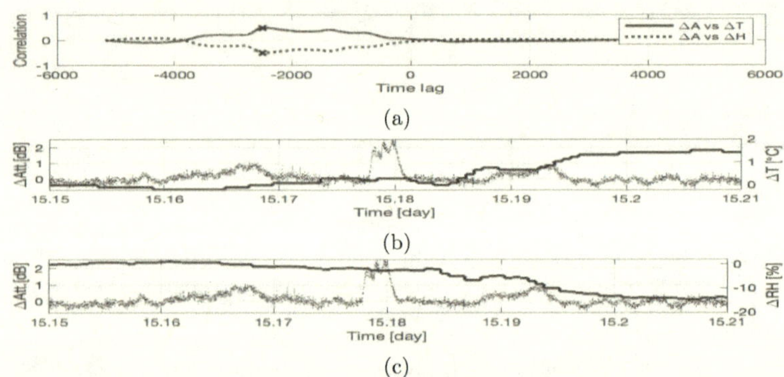

Figure 4.46: Poor correlation between attenuation and temperature/humidity, morning 16th of April.

Figure 4.47 show similarities in correlation function to a rain event but without a clear indication by the meteorological data. There are similarities to clear-sky regions which indicate no relationship and the source can thus not be identified.

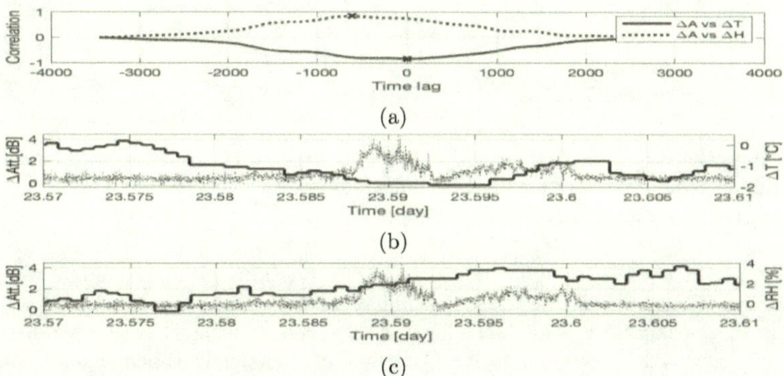

Figure 4.47: Symmetric correlation function with large correlation value. No clear relationship between temperature/humidity and attenuation. Similar to clear-sky region, evening 24th of April.

The same applies to Figure 4.48. There are an even higher attenuation above 5 dB but with low humidity increase. The temperature seem to drop in relationship to the attenuation.

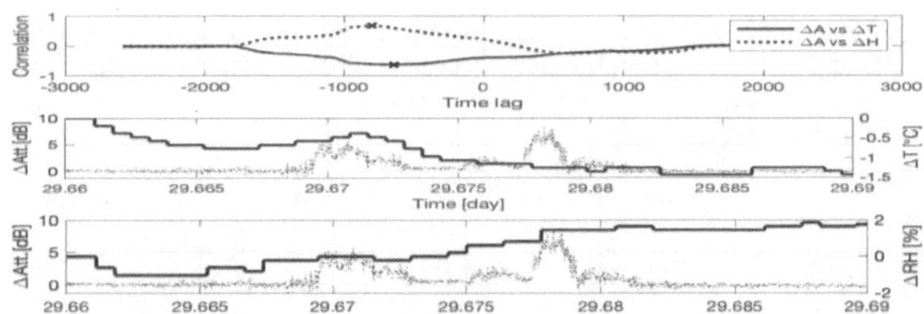

Figure 4.48: Symmetric correlation function shape with moderate correlation value. Humidity slowly increase while temperature show slight fluctuation with attenuation event, evening 30th of April.

Figure 4.49-Figure 4.52 show correlation functions with a seemingly random pattern. This is due to a larger independent behavior from both temperature and humidity. This behavior was also seen for some clear-sky regions.

Figure 4.49 show no indication of rain with independently changing temperature and humidity.

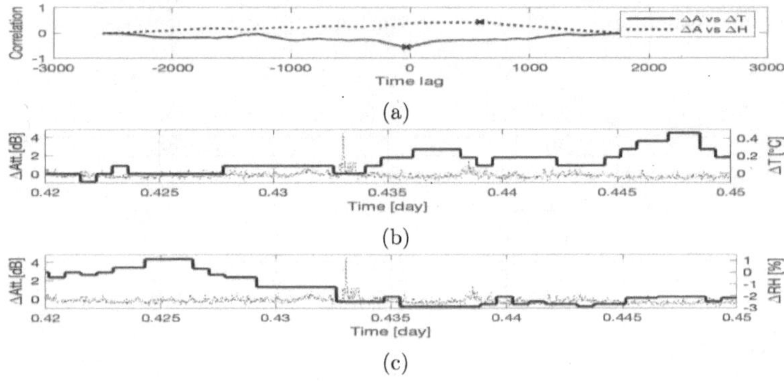

Figure 4.49: Low correlation and asymmetric correlation function shape, before noon 1st of April.

Figure 4.50 show an attenuation event active over a longer period of time with both increasing temperature and increasing humidity. The morning region would show natural decreasing humidity while temperature rise. There is thus a influence on the humidity as it is seen increasing from time 6,4 until near the end. Since there is no drop in temperature this region could potentially show attenuation from cloud and atmospheric gases.

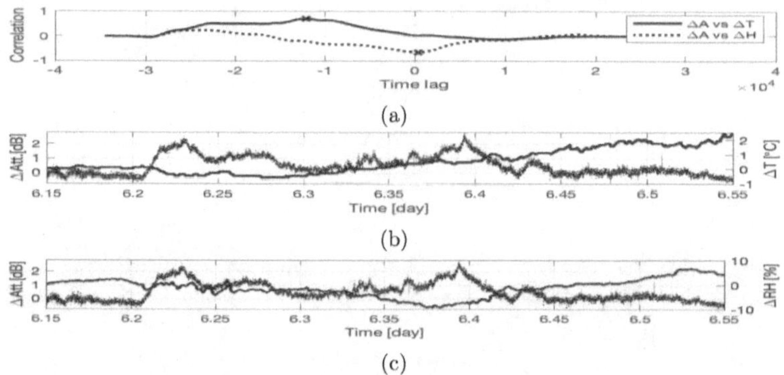

Figure 4.50: Moderate correlation and asymmetric correlation function shape. No clear relationship between attenuation and temperature/humidity, before noon 7th of April.

Figure 4.51 have fairly constant levels of temperature/humidity which yields the low correlation. There might be a relationship at the end of the region after time 28,775 where the temperature decrease and humidity increase. This could be a small rain attenuation event but could also be natural occurring behavior since it is evening.

Figure 4.52 show a steady decrease for humidity while temperature show independent fluctuations. The decrease of humidity suggest other source than rain.

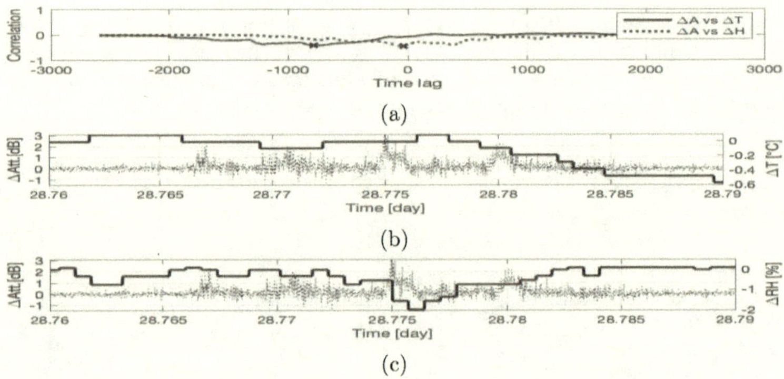

Figure 4.51: Very low correlation value and asymmetric correlation function shape. Both meteorological data show little to no change, evening 29th of April.

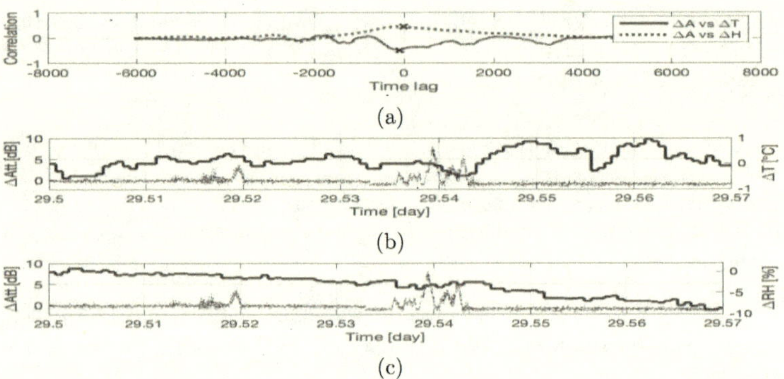

Figure 4.52: Moderate correlation value and asymmetric correlation function shape. Temperature fluctuate while humidity show steady decline, noon 30th of April.

Some of the regions presented in this section showed a different behavior from rain. Only Figure 4.45 showed clear rain characteristics. Small attenuation events that occur during evening with similarities of rain events are not possible distinguish the source. This was shown by Figure 4.44, Figure 4.47 Figure 4.48, and Figure 4.51. There were four figures that had no similarities to rain events and is most likely due to cloud and atmospheric gases (see Figure 4.46, Figure 4.49, Figure 4.50, Figure 4.52).

Characteristics of cloud and atmospheric gases.

- Asymmetric correlation function between attenuation/shifted temperature and attenuation/shifted humidity.
- Decreasing humidity and/or increasing temperature in relationship to attenuation event.
- Attenuation event without any strong correlation to surface meteorological data

Chapter 5

Conclusion

There are several atmospheric effects that can attenuate the signal at millimeter wavelength. The main sources of attenuation being rain, cloud, and atmospheric gases. Analysis on the transmitted Alphasat signal with frequency 39,402 GHz over the month of April 2018 indicated large attenuation event in relationship to rain rate. The surface meteorological data, temperature and humidity indicated a strong relationship during rain events. There was in total seven different time periods over the month with indicated rain rate (see Figure 4.1). Six of these regions showed a similar behavior in how the temperature and humidity responded to the increased attenuation. We saw an increase in humidity and decrease in temperature with mirror-like behavior as indicated by Figure 4.20 (b)-(g). We exclude the first rain event due to it being a remnant of a previous rain event from earlier month. Even with a small sample size of six full scale rain events the conclusion is clear: surface meteorological data is a great indicator for rain attenuation. This behavior was also found for an attenuation event without indication by rain rate. Figure 4.45 showed strong rain characteristic in the behavior of temperature and humidity in relationship to the attenuation. Many other regions e.g. Figure 4.44, Figure 4.47, Figure 4.48, and Figure 4.51 also show this behavior but due to the small attenuation and occurring during evening it could not be distinguishable from the natural change due to diurnal cycle.
The surface meteorological data over clear-sky regions i.e. regions with minimum attenuation gave information about normal occurring behavior without relationship to attenuation events. It is useful to study the mean change of meteorological data over regions to get an idea of the normal behavior over a month. This would show if certain behaviors found in attenuation regions were unique to that event or not. This gave an insight for events that otherwise could be misled as rain as previously described above. It also show normal fluctuations in the parameters and what can be expected in normal circumstances.
There are many small attenuation events on the signal over the course of a month. To analyse every single region is a tedious task and the reason why attenuation boundaries were set to only look at the most significant regions. The filtering function looked for attenuation events over 2,5 dB which

provided 16 regions to study closer. Below this point the amount of events increased quickly. As listed in Table 4.1, a number for 53 regions were found between 1,5 dB and 2,5 dB. For cloud and atmospheric gases the relevant attenuation levels occur in these regions. The prediction methods showed that the attenuation from these effects should not exceed around 3 dB.
There were essentially four regions (Figure 4.46, Figure 4.49, Figure 4.50, Figure 4.52) above 2,5 dB that showed a characteristic that does not correspond with rain characteristics. Four other regions had a indistinguishable source which could be from either rain or other atmospheric effects. The characteristic found for cloud and atmospheric gases were an asymmetric correlation function between attenuation/temperature and attenuation/humidity. A clear distinction from rain event is an increase of temperature and/or decrease in humidity. There no clear clear characteristic for these type of regions other than being different from rain characteristics. A larger quantity of regions could improve this aspect. There are many lower attenuation event but also increase the difficulty in identifying its source. This is also difficult since the clear-sky level is not exact (calculated as the mean in this study) which interfere at these small decibel events. Attenuation events occurring during afternoon until early morning could very well show rain characteristic due to the temperature/humidity diurnal cycle.

Does the surface meteorological data indicate attenuation by cloud and atmospheric gases?

There were no clear behavior in meteorological data over the attenuation events potentially caused by cloud and atmospheric gases. However, the rain events had such distinct behavior which can be used as a benchmark. The final conclusion is thus that the surface meteorological can be used for its poor indication to find possible attenuation events by atmospheric effects other than rain. Lower attenuation levels increase the difficulty in identifying its source.

5.1 Future work

There are improvements that can be done in both practical and theoretical work. A larger quantity of meteorological sensors could improve the accuracy e.g. using rain rate sensors on the surface of the Earth which follows the signal path. This would allow the indication from rain events that does not travel directly over the antenna. The largest improvement for other atmospheric effects than rain would be to gather the meteorological data over the vertical column. This could be done by daily measurements with radiosondes or by a radiometer. This would also require the most resources and is the reason why simple approximation methods are being developed as is it the most practical. The ITU method for predicting attenuation by atmospheric gases required the barometric pressure which were not locally available. This could perhaps increase the accuracy of the prediction model slightly.
Improvements could also be made in pattern recognition software to easier distinguish regions with a certain behavior. The observed behavior found here could be implemented to find more similar regions in the lower attenuation events (below 2,5 dB). This would increase the quantity and improve the conclusions whether cloud and atmospheric gases can be indicated by surface meteorological data.

Appendices

Appendix A

Rain regions

A.1 Cross-correlation: attenuation/shifted temperature, attenuation/shifted humidity

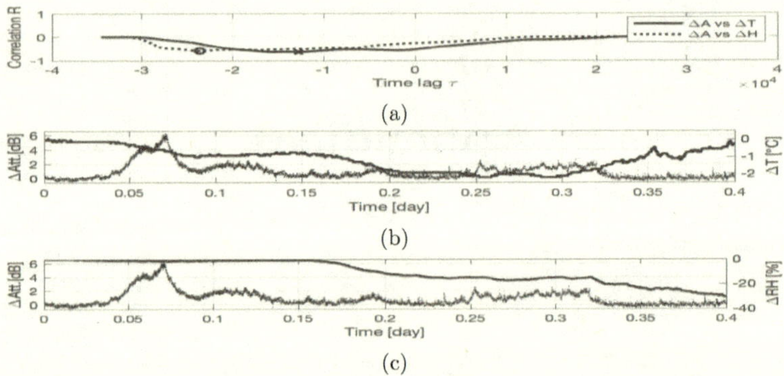

Figure A.1: Both decreasing temperature and humidty. High hymidity drop suggest remnant of larger rain event from previous month, 1st of April.

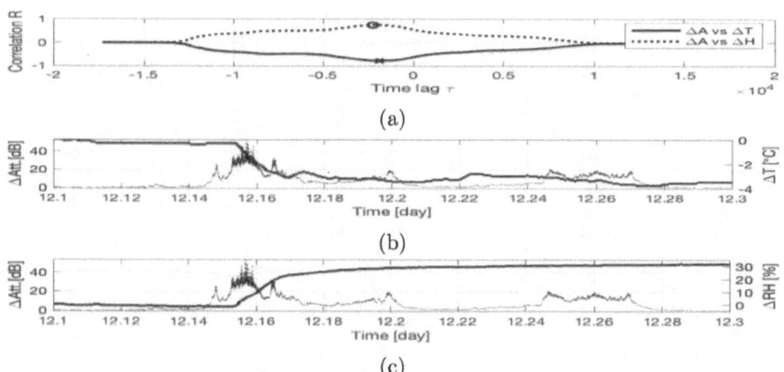

Figure A.2: Distinct decrease of temperature and increase of humidity in relation to high attenuation. Typical rain characteristics, 13th of April.

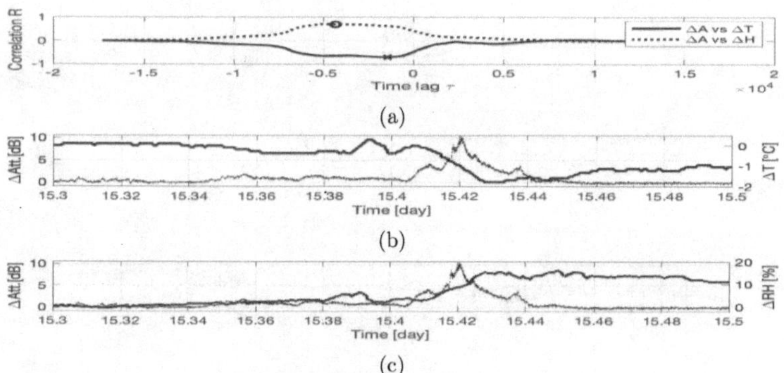

Figure A.3: Distinct decrease of temperature and increase of humidity in relation to high attenuation. Typical rain characteristics, 16th of April.

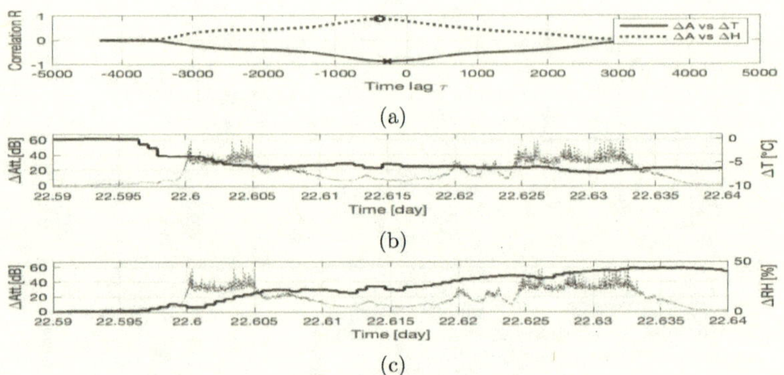

(a)

(b)

(c)

Figure A.4: Distinct decrease of temperature and increase of humidity in relation to high attenuation. Typical rain characteristics, 23rd of April.

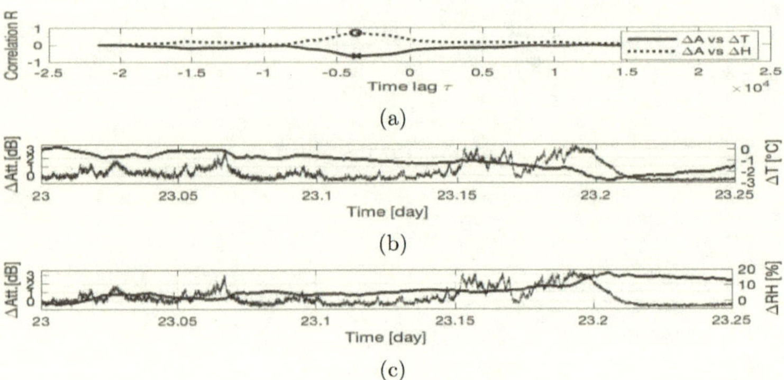

(a)

(b)

(c)

Figure A.5: Small decrease of temperature and increase of humidity in relation to high attenuation. Strong relationship between temperature and humidity suggest rain event, 24th of April.

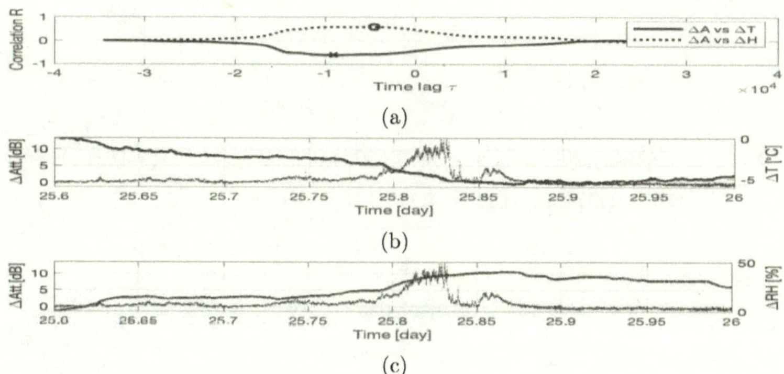

Figure A.6: No clear decrease in temperature or increase in humidity in relation to attenuation event. Coincide with diurnal cycle which natural show same behavior of temperature and humidity. High humidity levels suggest rain event, 26th of April.

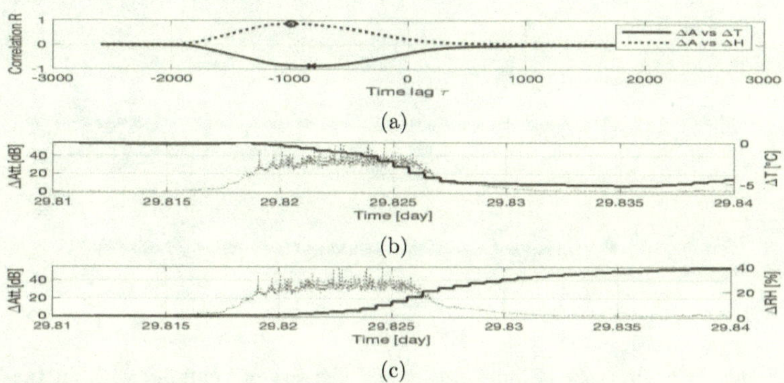

Figure A.7: Distinct decrease of temperature and increase of humidity in relation to high attenuation. Typical rain characteristics, 30th of April.

Appendix B

Clear-sky regions

B.1 Comparison: attenuation/temperature, attenuation/humidity

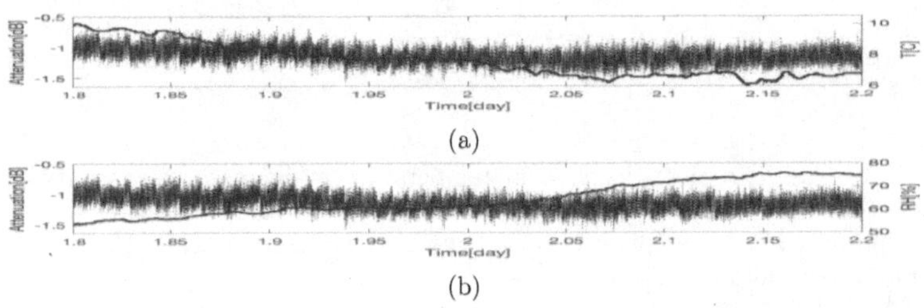

(a)

(b)

Figure B.1: Stable decrease of temperature and increase of humidity, midnight 2nd of April.

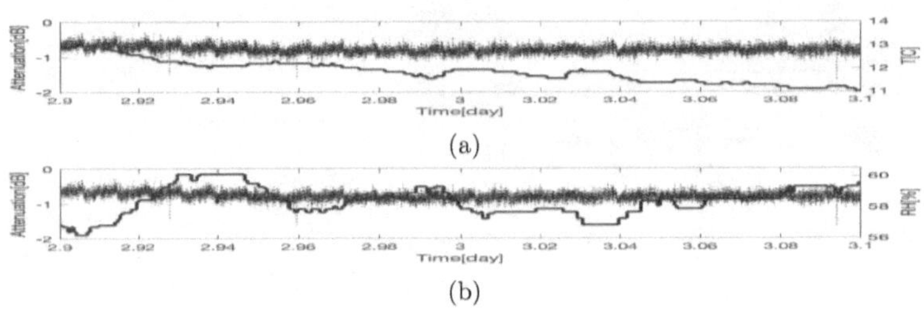

(a)

(b)

Figure B.2: Decrease of temperature and increase of humidity with fluctuations, midnight 3rd of April.

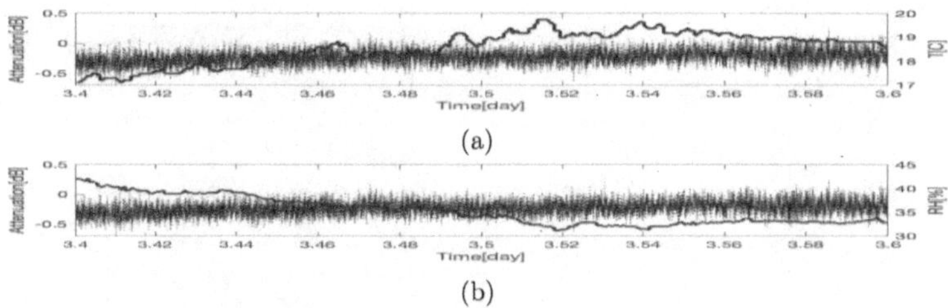

(a)

(b)

Figure B.3: Increase of temperature with small fluctuations with decrease of humidity, noon 4th of April.

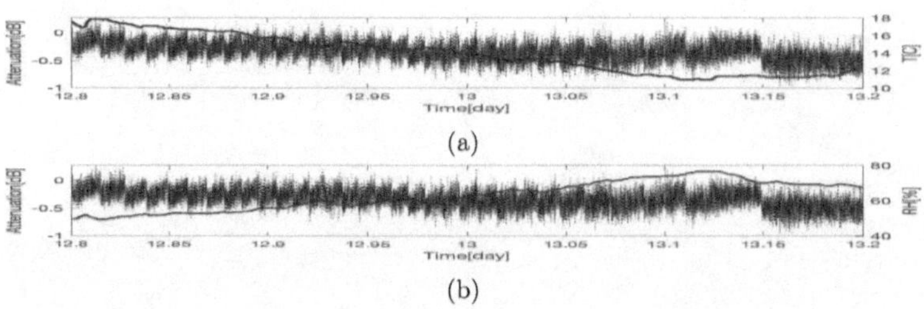

(a)

(b)

Figure B.4: Decrease of temperature and increase of humidity, midnight 13th of April.

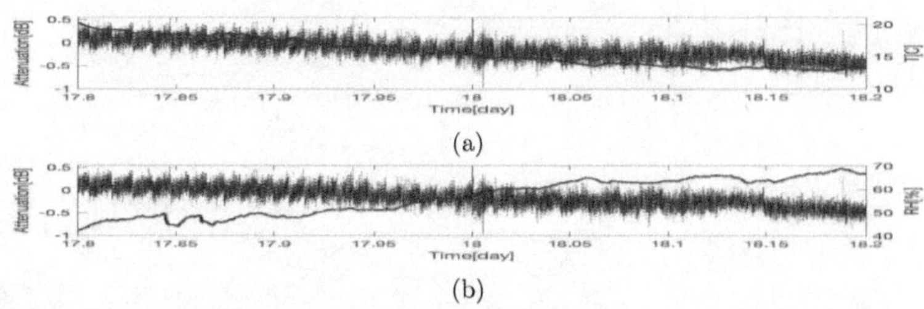

(a)

(b)

Figure B.5: Decrease of temperature and increase of humidity with fluctuations, midnight 18th of April.

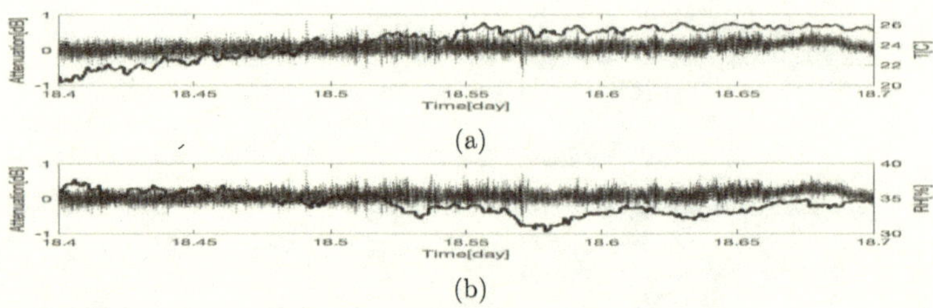

(a)

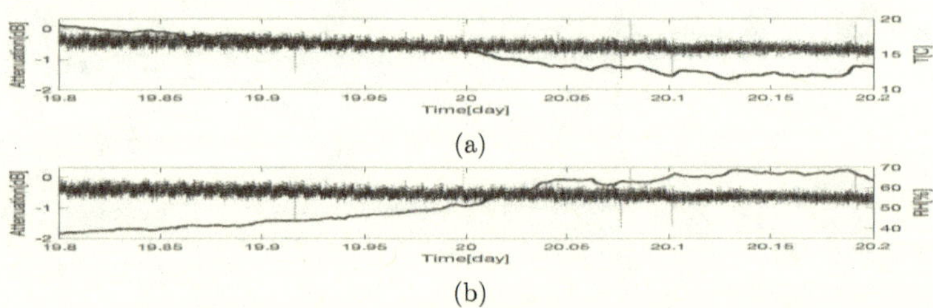

(b)

Figure B.6: Increase of temperature with small fluctuations and humidity fluctuating around 35%, noon 19th of April.

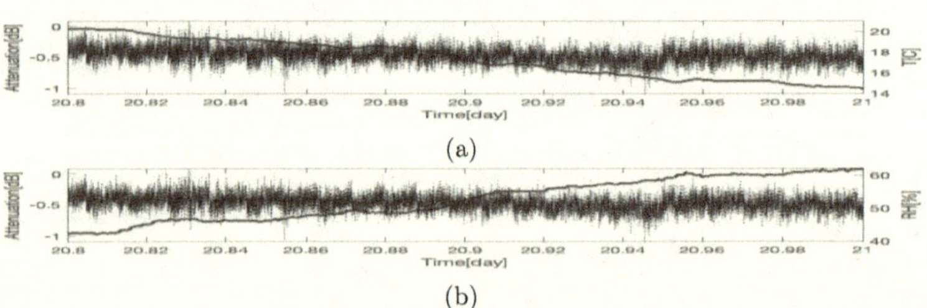

(a)

(b)

Figure B.7: Decrease of temperature and increase of humidity, midnight 20th of April.

(a)

(b)

Figure B.8: Decrease of temperature and increase of humidity, before midnight 21st of April.

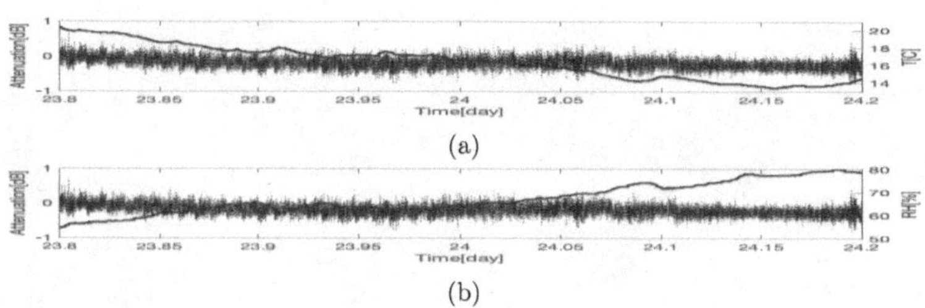

Figure B.9: Decrease of temperature and increase of humidity, midnight 23rd of April.

B.2 Cross-correlation: temperature/shifted humidity

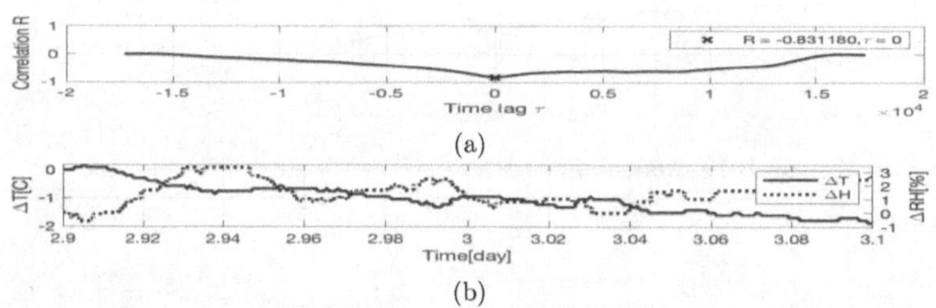

Figure B.10: Maximum correlation R=-0,83 at $\tau = 0$ time lag.

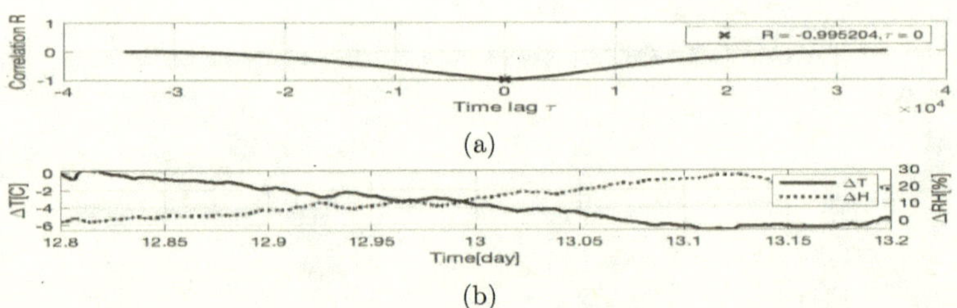

(a)

(b)

Figure B.11: Maximum correlation R=-0,99 at $\tau = 0$ time lag.

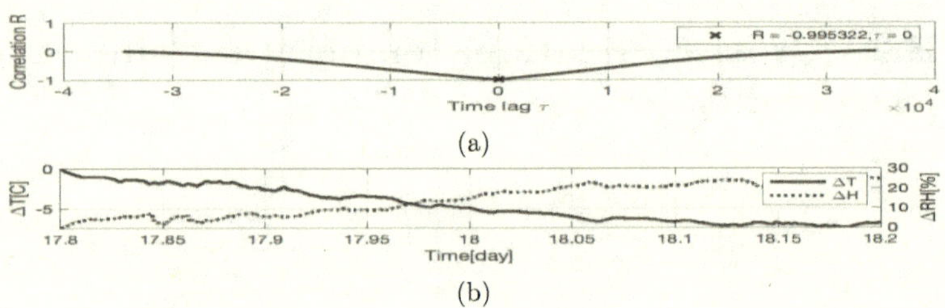

(a)

(b)

Figure B.12: Maximum correlation R=-0,99 at $\tau = 0$ time lag.

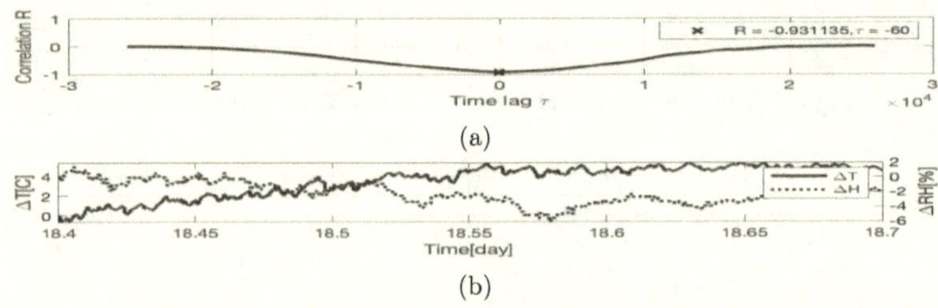

(a)

(b)

Figure B.13: Maximum correlation R=-0,93 at $\tau = -60$ time lag.

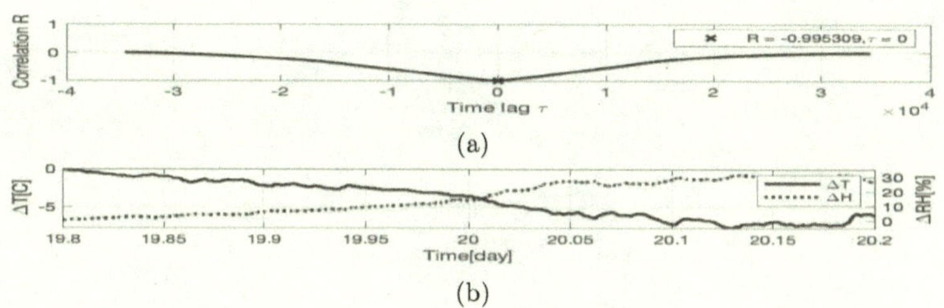

Figure B.14: Maximum correlation R=-0,99 at $\tau = 0$ time lag.

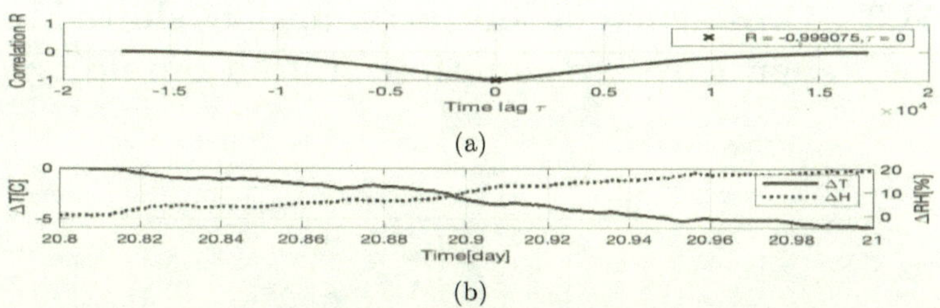

Figure B.15: Maximum correlation R=-0,99 at $\tau = 0$ time lag.

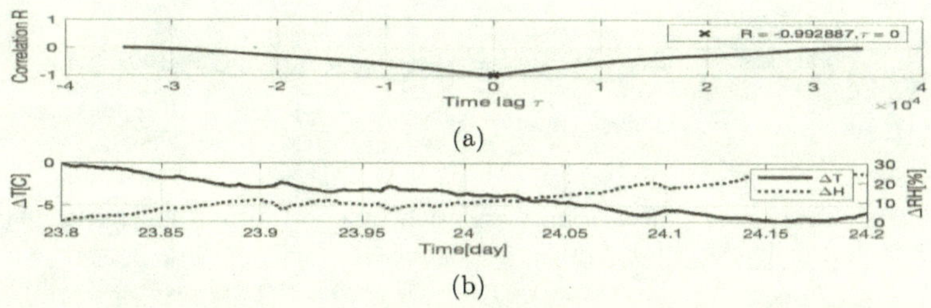

Figure B.16: Maximum correlation R=-0,99 at $\tau = 0$ time lag.

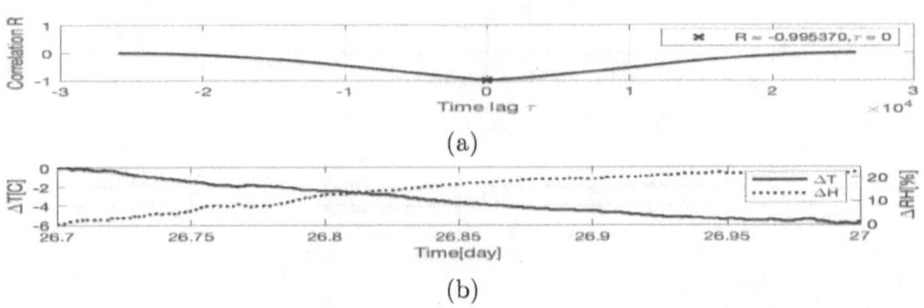

Figure B.17: Maximum correlation R=-0,99 at $\tau = 0$ time lag.

B.3 Cross-correlation: attenuation/shifted temperature, attenuation/shifted humidity

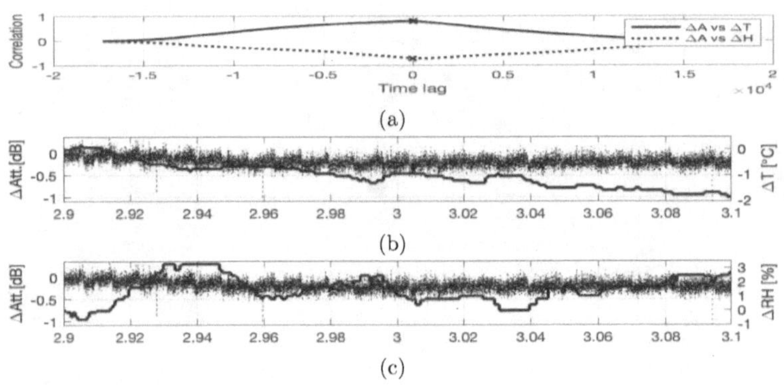

Figure B.18: Slowly decreasing correlation function with symmetric shape Fig. (a), midnight 3rd of April.

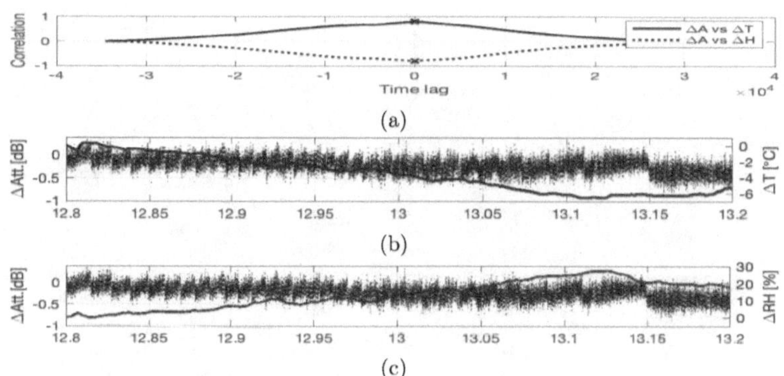

Figure B.19: Slowly decreasing correlation function with symmetric shape Fig. (a), midnight 13th of April.

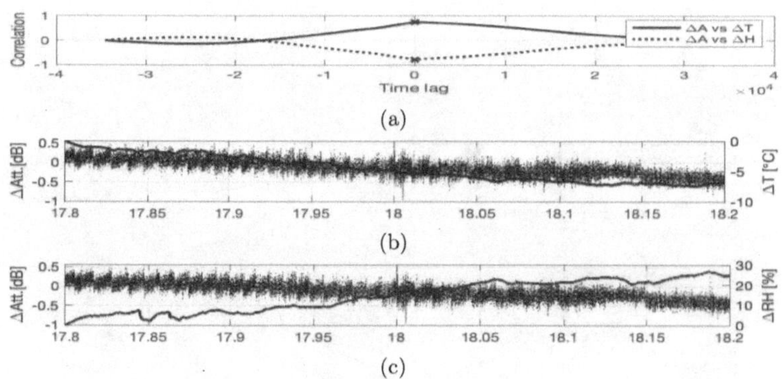

Figure B.20: Slowly decreasing correlation function with symmetric shape Fig. (a), midnight 18th of April.

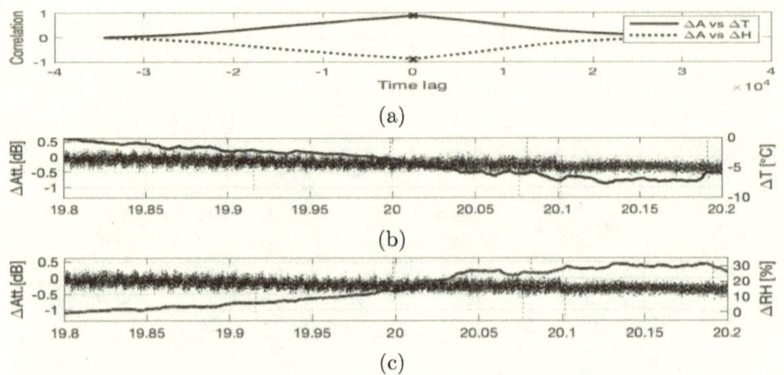

Figure B.21: Slowly decreasing correlation function with symmetric shape Fig. (a), midnight 20th of April.

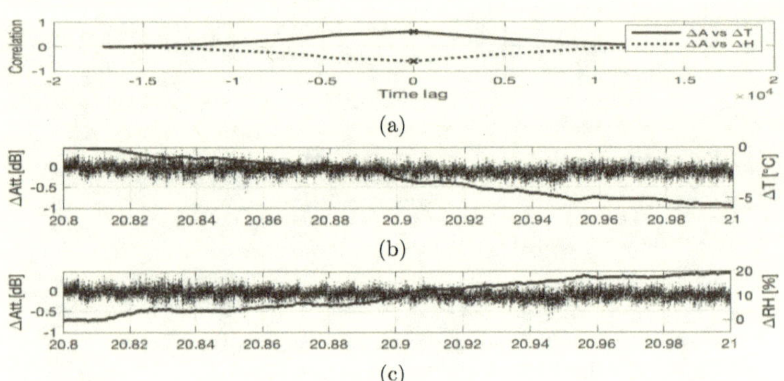

Figure B.22: Slowly decreasing correlation function with symmetric shape Fig. (a), before midnight 21st of April.

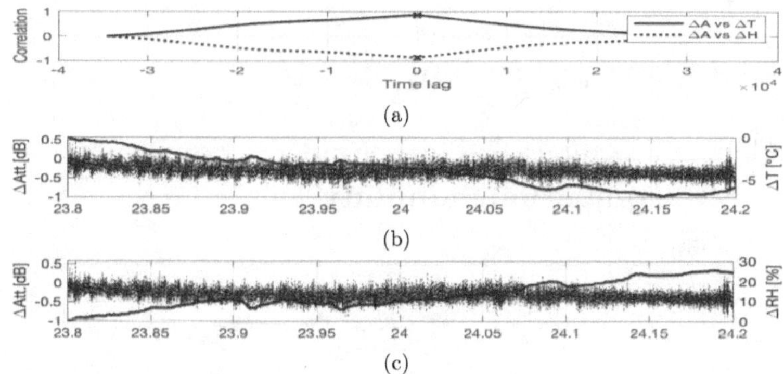

Figure B.23: Slowly decreasing correlation function with symmetric shape Fig. (a), midnight 24th of April.

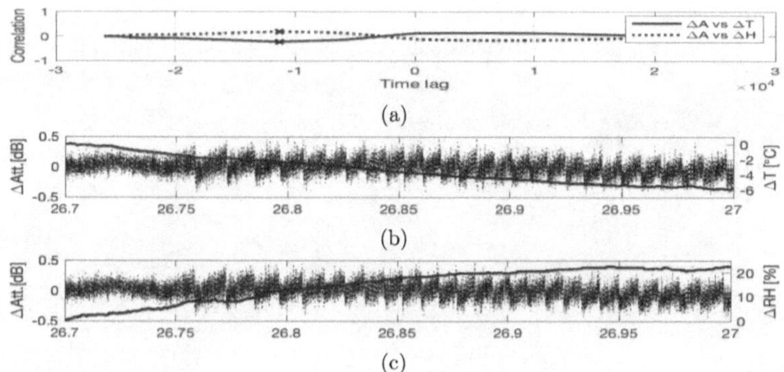

Figure B.24: Low correlation function with symmetric shape Fig. (a), before midnight 27th of April. This behavior is due to the tracking system adjusting the direction of antenna as pointing error reaches +/- 0,5 dB.

Appendix C

Cloud and gas regions

C.1 Comparison: attenuation/temperature, attenuation/humidity

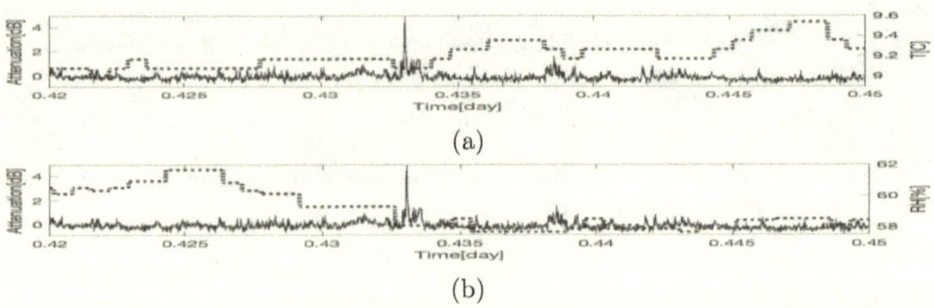

(a)

(b)

Figure C.1: 4 dB attenuation event before noon on 1st of April.

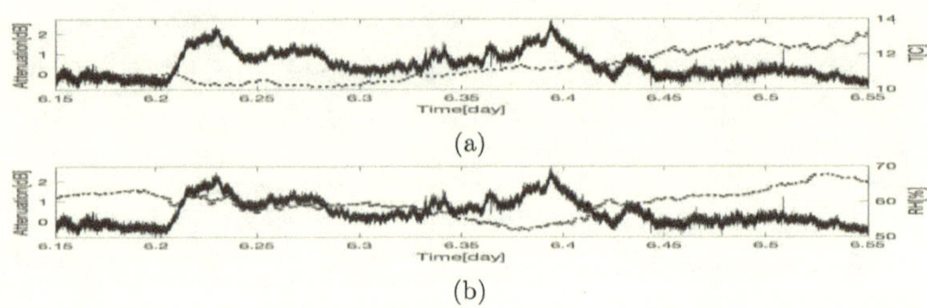

(a)

(b)

Figure C.2: 2 dB attenuation event in the morning on 6th of April.

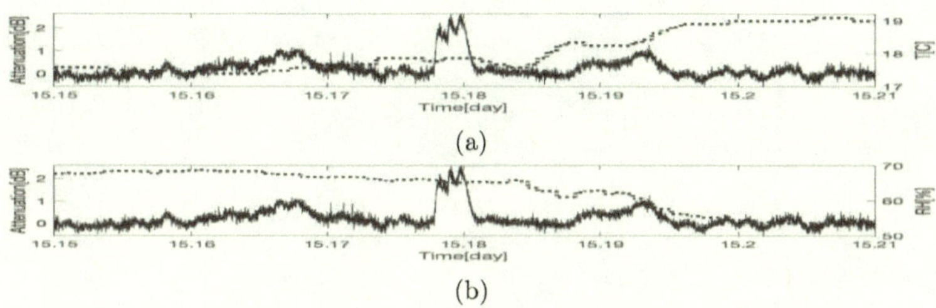

Figure C.3: 2 dB attenuation event in the morning on the 15th of April.

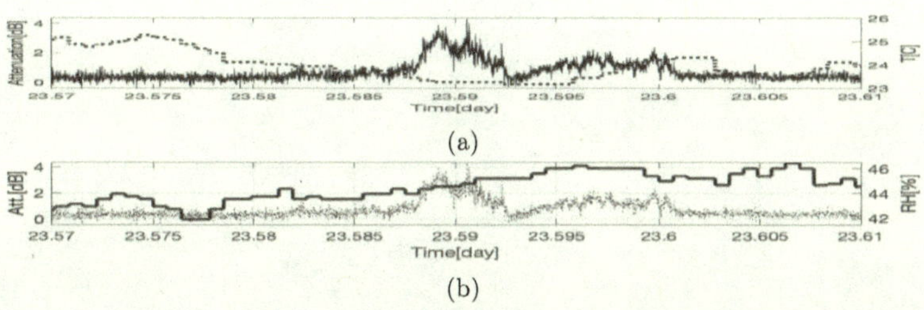

Figure C.4: 4 dB attenuation event after noon on the 24th of April.

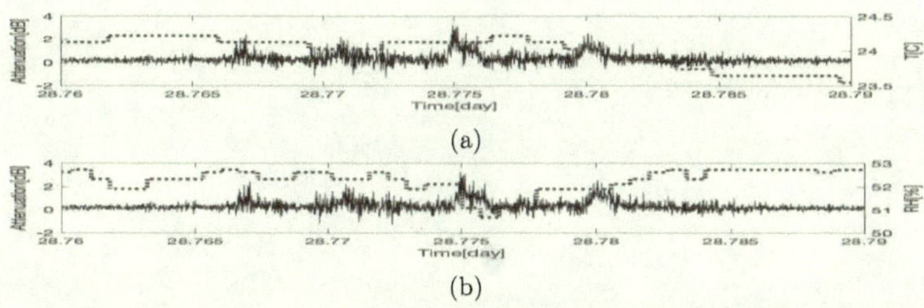

Figure C.5: 3 dB attenuation event in the evening on the 29th of April.

www.ingramcontent.com/pod-product-compliance
Lightning Source LLC
LaVergne TN
LVHW041127150826
845673LV00007B/2206

* 9 7 8 3 3 8 4 2 7 6 6 3 6 *